Chinese Painting Colors

中國畫顏色的研究

于非闇著

Chinese Painting Colors

Studies of Their Preparation and Application in Traditional and Modern Times

by Yu Feian

Translated by
Jerome Silbergeld and Amy McNair

Hong Kong University Press
Hong Kong

University of Washington Press
Seattle and London

Published in the United States of America by the University of Washington Press
and in Hong Kong by the Hong Kong University Press

Copyright © 1988 by the University of Washington Press
Printed in Hong Kong

Library of Congress Cataloging-in-Publication Data

Yü, Fei-an.
  Chinese painting colors.

  Translation of: Chung-kuo hua yen se ti yen chiu.
  Bibliography: p.
  Includes index.
  1. Colors. 2. Color in art. 3. Painting, Chinese—
Technique. I. Silbergeld, Jerome. II. McNair, Amy.
III. Title.
ND1510.Y813 1988   751.42'5   85-24638

ISBN 0-295-96356-5   (University of Washington Press)
ISBN 962-209-222-5   (Hong Kong University Press)

Contents

Color Chart of Traditional Chinese Painting Pigments follows page 34

Chronology of Chinese Dynasties

Shang   ca. 1600 B.C.– ca. 1100/1050 B.C.
Zhou   ca. 1100/1050 B.C.–221 B.C.
Qin   221–206 B.C.
Han   206 B.C.–221 A.D.
  Western Han   206 B.C.–9 A.D.
  Eastern Han   25 A.D.–221 A.D.
Three Kingdoms   220–265
  Wei   220–265
  Wu   222–258
  Shu-Han   221–296
Western Jin   265–317
Northern and Southern Dynasties   265–581

(Southern Dynasties)	(Northern Dynasties)
Eastern Jin 317–420	Northern Wei 383–534
Liu Song 420–479	Eastern Wei 534–549
Southern Qi 479–502	Western Wei 535–556
Liang 502–557	Northern Qi 550–557
Chen 557–589	Northern Zhou 557–581

Sui   581–618
Tang   618–907
Five Dynasties   907–960
  Later Liang   907–923
  Later Tang   923–936
  Later Jin   936–947
  Later Han   947–951
  Later Zhou   951–960

Song   960–1279   (Northern Dynasties)
  Northern Song   960–1127   Liao   907–1125
  Southern Song   1127–1279   Jin   1115–1234
Yuan   1279–1368
Ming   1368–1644
Qing   1644–1911

Translators' Introduction

Very little has been written about the use of color in Chinese painting. Over the centuries, many artists and writers about the art of painting cultivated an unspoken prejudice against any prominent use of, or even discussion of, color in painting. The intellectual outlook of the literati class helped shape this aesthetic sensibility, which was derived from the great value they laid on detachment from worldly things and the stilling of human emotions, on harmony and simplicity. Historically, this outlook was already apparent in the early stages of traditional Chinese philosophy, in which aesthetics was securely linked to the constraints of Chinese ethics. Confucius himself helped set this severe tone, lamenting that he had not yet met any man who loved virtue as much as he loved beauty. A precursor of the literati attitudes toward color in centuries to come may be found in Confucius' comments on court robes, where he claimed that the superior man would never wear violets or purples, in other words flamboyant, intermediate colors, rather than pure colors (the five pure colors being red, yellow, blue-green, black, and white); he set a still stricter standard for himself, refusing even in informal dress to wear anything but white, yellow, or black.[1] In a similar vein, the early Daoist classic, the *Dao de jing,* described mankind as crazed by its love of beauty and pleasure and warned that "the five colors will blind the eye" to true perception.[2]

Even the Chinese language reinforced this negative attitude toward color, as the written character for color, *se* 色 , carried the ancillary meanings of "beauty," "passion," "anger," and "lewdness"; when Confucius lamented the love of beauty, he used this word, and elsewhere, when he enumerated the three things to be guarded against by the superior man, the first was the young man's desire or lust, *se*.[3] Such values colored—or discolored—the art of Chinese painting. Many Chinese painters, particularly those amateur artists of the elite scholar-class who consciously applied such ethical precepts and constraints, avoided the superficial and shunned the use of color, pursuing instead linear structure or "bone," which was thought of as more in accord with moral integrity and inner principles.

A social component to this aesthetic should also be observed, namely the disdain with which the scholar-class viewed manual labor and technical skill. As painting pigments were not available ready-made for purchase and use before the eighteenth century, many amateur scholar-painters must have regarded as distasteful the

[1]James Legge, trans., *The Chinese Classics, I: The Confucian Analects* (reissued Hong Kong: Hong Kong University Press, 1960), pp. 222, 298, 230–31.

[2]Cf. Ch'u Ta-kao, trans., *Tao Te Ching* (London: George Allen and Unwin, 1937), p. 24.

[3]Legge, *Chinese Classics, Analects,* pp. 222, 298, 312–13.

time-consuming preparation of pigments and the technical skill needed for their proper application. It is not surprising that in the gradual evolution of a "scholar's style" of painting during the late-Tang, Song, and Yuan dynasties color should have come to play a lesser role in Chinese painting than it had in earlier periods, when the art had been dominated by technically skilled professionals. Furthermore, it is not surprising that these scholars, who monopolized the writing of art history, art theory, and criticism, should have found it similarly displeasing to discourse on the techniques of painting in color. As a result, the written record is even more prejudicial in its treatment of color than the actual history of usage; a comparison of the written and visual records suggests that the role of color in Chinese painting history has been poorly represented by writings on the subject.

In the history of Chinese painting, one of the earliest and most enduring of assumptions was that painting consisted of first providing a linear structure, usually done in black ink, a fact conditioned by the nature of the Chinese brush being more compatible with linear movement than with the covering of broad areas. Only subsequently were these outlines filled in with washes of color, so that color did not play the major structural role it did in Western painting. Yet in early times, the design value of color was rarely overlooked, and although some sketch-like works in ink alone existed, they were hardly typical. As the manufacture and techniques of applying color matured, an increasingly sophisticated use of color became indispensable to the artist. A proper term for "painting" was "red and blue" or "the reds and the blues" (*dan qing*). By the early Han period, bright colors were often highly esteemed, although restraint in their application was also to be found. In the Six Dynasties and Tang periods, exotic color schemes, some of foreign origin, were very fashionable, and the jewel-like "blue-and-green" land-scapes, painted in bright mineral pigments, were thought of in later periods as the foremost mark of painting at that time.

Only with the passage from youth to maturity after the early Tang period was the Chinese painting tradition marked by the conscious suppression of the use of color, at least in some quarters. Monochrome ink painting first gained adherents in the eighth and ninth centuries and rapidly began to change the nature of Chinese art. And yet while some pioneers of monochrome painting placed foremost emphasis on brush and ink textures, many other artists simply shifted their interest from color hues to ink tones ("the colors of ink"), a new standard which Chinese critics were quick to grasp. In 847, Zhang Yanyuan wrote in his monumental history of Chinese painting, *Record of the Famous Painters of the Successive Dynasties*: "It may be said, one's ideal is attained if, in spreading the ink, the five colors are there complete."[4] From then on, ink and its tonality replaced pigments and hues in most writer's idiom and dominated the subsequent writings on Chinese painting.

But in painting, all attention did not forever turn away from the subtleties of hue. In the Song Dynasty, even as monochrome painting brought Chinese art to one of its greatest moments, "boneless" (non-linear) paintings in colored pigments were also produced that represented a peak of technical perfection in the representational use of colors, the finest results being marvelously naturalistic. And even as the scholar-amateurs of the Yuan gave visual definition to a "scholar's style" of painting, the perpetuation and refinement of Song styles by more conservative

[4]Cf. William Acker, trans., *Some T'ang and Pre-T'ang Texts on Chinese Painting*, 2 vols. (Leyden: E. J. Brill, 1954, 1974), 1:185.

artists, like the bamboo-painter Li Kan, and the historically conscious revival of pre-Song styles by landscapists like Qian Xuan and Zhao Mengfu kept the use of color alive and creative in a time of diminished influence. In the early Ming period, academic revivalism brought Song color-naturalism back, especially in the production of bird-and-flower paintings. More significantly, with the decline and subsequent revival of "scholar's-style" painting in the fifteenth century, color came to play a new and broadened role in the art of the scholar-amateurs, even though this went unheralded in the scholarly literature. Although that literature continued to speak only of "ink-play," scholar-painters came to develop a highly sensitive "color-play" with delicate washes of pigment, usually with vegetable pigments rather than the more opaque mineral pigments dominant in earlier times. Many major scholar-artists of later times, from Wen Zhengming in the early sixteenth century to Daoji in the late seventeenth century, were masters of hue as well as tone, complementing their preeminent brush rhythms with a refined and creative use of pigments. One needs only to compare their original works with black-and-white illustrations of them to see how badly such paintings fare when deprived of their hues. Throughout the last hundred years and down to the present day, under the combined influence of traditional masters like Daoji, folk aesthetics, and the importation of brighter Western pigments, Chinese paintings have come even more greatly to emphasize strong, bright colors.

Thus, the disparity between the role of color in Chinese painting and the scant attention it has received in writing is great. In our opinion, this book represents the most important contribution yet made to the redress of that disparity. Written by a prominent painter of flowers and birds, Yu Feian 于非闇 (1889-1959; his proper name was Yu Zhao 于照), under the title *Studies of Chinese Painting Colors* 中國畫顏色的研究, it was originally published in Beijing in 1955 by the Zhaoyang Book Company; it went into a second printing in 1957 and still remains popular in China today. (We have taken the liberty of issuing this translation under a modified and more fully descriptive title.) Yu's book is primarily technical in nature, concerned with the selection, preparation, and application of materials, addressing as well such issues as the sizing of silks and paper. It is not greatly concerned with style or the aesthetics of color composition, but we hope that the appearance of this work in translation will facilitate greater appreciation and stimulate deeper inquiry into this aspect of Chinese painting.

It is not surprising that such a work first appeared in China in the 1950s, coming with the release in the early and mid-twentieth century of writers and artists from centuries of inherited restraints and with a new emphasis on progress and scientific inquiry. In addition to the wealth of technical information offered here, Yu Feian presents an important perspective on Chinese painters and how they worked. While the Chinese painting process is often viewed as a loosely articulated, relaxed, and frequently amateurish activity, the artistic processes revealed in this book are practical, intensely physical, and even quasi-scientific. Perhaps such intensive activity best describes those artists of earlier times—the professional and academic painters—who often collaborated and divided their work into subspecialties, in contrast to the leisured scholar-amateurs who surely left some of the manual labor to their ever-present boy-servants. But just as surely, this activist view of the Chinese painter applied in some degree to all artists who prepared their own materials, and before the past two centuries this included most painters regardless of social background or financial means.

Yu Feian himself had much in common with the artists of earlier times. His favorite artists were academic painters of the Song period whose colorful, naturalistically depicted, and technically refined flowers and birds he took as his primary models and whose inquiring outlook inspired his own. While his own art was not limited to any one style, it was his conservative, richly colored painting style that secured for him a nationwide reputation and led him into the technical studies of painting colors that are presented in this work. The frontispiece of this book (reproduced from the cover of the original Chinese edition) is done in his own calligraphy, in the style of Song Emperor Huizong (Zhao Ji), the most prominent imperial patron and practitioner of this style of painting.

Yu Feian was a highly energetic individual, active on many fronts. After 1949, he served as a representative to the First and Second People's Congresses, as Director of the Beijing Municipal Literary Association's Standing Committee, as Vice-chairman of the National Painting Symposium in Beijing, and at the time of his death in 1959 was Vice-director of the National Painting Academy in Beijing (now the Central Academy of Fine Arts), which had been established just two years earlier. Yu's book is imbued with the best elements of the intellectual spirit of its time, not without its wooden political jargon (which we have not attempted to eradicate) but inspired by the urge to examine the past for the best that it had to offer to a modern China, to be scientific in this examination, and to broaden the social scope of the examination by including folk artists and folk traditions of painting alongside paintings in the more aristocratic tradition.

In order to indicate something more of the spirit which gave rise to this book we would like to supplement our introduction by quoting liberally from an obituary written by one of his colleagues at the National Painting Academy, Wang Youshi 王友石 in *Meishu (Fine Arts Magazine)*.[5]

His whole life long, Mr. Yu devoted himself to the occupation of traditional painting. In the old society [before 1949], he encountered bitter difficulties, working all day long to support his hungry family, even to the point, in the most difficult of times, of selling all his painting materials in order to preserve some minimal existence, and enduring such conditions until Liberation came to Beijing. But even in such circumstances, Mr. Yu persevered in his research into traditional painting; working according to his own fixed schedule of daily investigations, he studied tirelessly, actually becoming stronger with age, with a spirit of adhering to principle in the midst of hardships.

Mr. Yu's paintings, which explored many types and varieties of form, included much rich coloration and went through a long period of rigorous study that added substance to his own personal style. He used to ask about painting methods from the older folk artists, even ... [asking about painting methods from] the elderly Qi Baishi [d. 1957]. He studied the freehand painting of flowers and birds and the freehand painting of landscapes, and he painted decorative patterns. His methods were not limited to a single type, for he sometimes used the plain-line method, sometimes the boneless method, sometimes the method of drawing the flowers and dotting in the leaves, sometimes the method of drawing and then adding color, and sometimes the method of unmodulated outlines with flat wash. Before he came gradually to develop his own style, his technique depended on copies made after the old masters. From 1939 on, when he began his studies of painting in detailed brushwork, he emphasized the study of the ancient masters' methods. What he revered most of all were the flower and bird painters of the Song. He said that

[5]Excerpted from *Meishu (Fine Arts Magazine)*, August 1959, pp. 39–40. For further references regarding Yu Feian, including sources for published reproductions of his works, see Ellen J. Laing, *An Index to Reproductions of Paintings by Twentieth-Century Chinese Artists* (Eugene: University of Oregon Asian Studies Program, 1984), pp. 477–79.

in the flowers and birds of the Song masters the form and spirit were both captured, creating a subtle sense of nature, and that this should be studied. He said that enjoying the finest flower and bird paintings of the Song masters was like listening to the singing of birds, like smelling the fragrance of flowers. Because of the degree of his esteem for the Song masters and the depth of his understanding, his copies of the Song masters' paintings, in both his brushwork and use of color, were so absolutely lifelike that they could be confused with the originals. Mr. Yu's copies after ancient paintings took considerable effort. He used all the various methods of copywork, sometimes sketching on transparent paper placed over the original painting, using the method of transfer drawing; sometimes he would draw up a design from the original painting with willow charcoal, using the method of direct copy; sometimes he would carefully scrutinize the original painting and afterwards imitate it, drawing on his subjective grasp of it, using the method of copy from memory. Occasionally he would copy a painting that he loved some four or five times in order to absorb its finer points. His depth of understanding of the principle, 'Transmitting and conveying [earlier models, through] copying and transcribing,' from the 'Six Principles' [of Xie He],[6] was an accomplishment that aided Mr. Yu greatly in learning the fine tradition of national painting.

Mr. Yu loved to do research on the fine art of classical literature and the principles of classical painting. He often said, 'All of our nation's classical literature had one thing in common, which was the development of plots that would bring forth new concepts, stripping away all the old in order to provide people with new perceptions.' For this reason, he adhered to the ancient realistic tradition that 'to study the ancients is no match for the study of nature,' and he then proceeded from doing copywork to becoming engaged in the activity and study of sketching from life. At that time, some artists only emphasized the spirit-resonance of brush and ink, forgetting forms and pursuing the inner essence, but Mr. Yu emphasized that form and spirit must both be captured. Without regard for the criticism of others, he persevered in the efforts he made in his own work, finally creating his own style. His techniques in painting flowers and birds carried on those of the Song period, while absorbing some of the finer aspects of folk art and the decorative arts and adding some of the knowledge he obtained from a lifelong practice of painstakingly observing real flowers and birds, so that he acquired a thorough mastery through the comprehensive study of his subjects and produced his own style.

His ability to attain this kind of accomplishment was related to the depth of his examination of the objects he painted. In the old society, although his financial means were insufficient, he usually found a way to raise flowers and birds, so that he could view and appreciate them and sketch them from life. He even made a several years' study of doves and published his *Notes From the Capital City on Raising Doves*. In order to examine the behavior of doves in flying and roosting, he used to climb up on the city walls to observe them, so as to inspect the various kinds of movements they made. Mr. Yu painted peonies over a long period of time and his paintings of them were numerous. His researches into the peony were numerous, his observation of them very meticulous. He paid attention to what kinds of peonies put forth what kinds of leaves. Whenever it was time for the peonies to flower, he would go to the public parks to watch the transformation of their blossoms. From the time their flowering began until the time it ceased, he would examine them from all points of view, painting the beautiful, artistic form of all of the blossoms to serve as material for the time when he would create a new painting sketchbook. He said that when the flowers began to bloom they should be painted immediately in a notebook of sketches from life, and that after they had flowered one should not be tardy in painting the leaves. As for painting the older stalks, they could most clearly be observed after the leaves had fallen. In painting the old stalks he used extremely firm brushstrokes, while he used supple brushstrokes for painting the leaves. In applying colors, he added repeated washes until they were very dense. As for the techniques of using color, he learned a great deal through study and published his *Studies of Chinese Painting Colors*.

Mr. Yu's great accomplishments were surely based on his careful study and painstaking practice and cannot be separated from his many-sided cultivation of the arts, including poetry and literature, calligraphy, seal-carving, opera, raising flowers and birds, and various other pursuits.

* * *

[6]James Cahill, trans., "The Six Laws and How To Read Them," *Ars Orientalis*, 4, p. 380.

Throughout our translation of this work, Yu Feian's original notes are indicated in the text by superscripted boldface letters and placed as endnotes following the text, while translators' notes are marked by numerals and appear as footnotes. In the text, comments in parentheses are those of the author, while translators' comments are placed in brackets. The present location of major paintings referred to in the text and published sources where readers may find colored illustrations of these paintings, if available, are listed in the Appendix. In the Index of this work, readers will find Chinese characters for painting colors and for technical terms used in the text, as well as those for Chinese artists, authors, and book titles cited in the text. In the text, book titles appear in translation whenever that provides greater meaning for the reader.

In translating this work, we were assisted by many scholars and painters more knowledgeable about painting techniques and materials than ourselves. Professor Jin Weinuo, Chairman of the Art History Department of the Central Academy of Fine Arts, Beijing, was most generous in his support of this project since its beginning and arranged for the production of the chart of traditional Chinese painting colors reproduced in this book. Mr. Jin Tian of the Palace Museum in Beijing, Department of Copying, prepared this chart (to which we have added four additional colors) with great skill and knowledge. Two Chinese painters made numerous corrections and helpful suggestions based on their personal experience: Mr. C. C. Wang of New York, a connoisseur, scholar, and artist whose paintings we both greatly admire, and Mrs. Yu Sung, Curator of Asian Art at the San Diego Museum of Art. Dr. John Winter, Conservation Scientist at the Freer Gallery of Art Technical Laboratory, provided considerable valuable information and recommendations regarding technical terms and procedures. Ellen Laing, Maude I. Kerns Professor of Oriental Art History at the University of Oregon, generously provided us with bibliographical resources on Yu Feian and also suggested many improvements in presentation. Professor James Cahill, University of California, Berkeley, and Professor Esther Jacobson, University of Oregon, both read and helped to improve the entire manuscript. To Professor Jacobson, as a former teacher of both of us, we would like to offer a special note of gratitude. To all of these individuals we extend our deepest appreciation.

Finally, we would also like to express our sincere gratitude to those programs at the University of Washington in Seattle whose financial assistance helped make this publication possible, including the Graduate School Research Fund; the Chester A. Fritz Endowment of the China Program, Henry M. Jackson School of International Studies; the School of Art; and the Division of Art History.

Jerome Silbergeld
University of Washington

Amy McNair
University of Chicago

Chinese Painting Colors

Chapter 1

The Varieties and Nature
of Chinese Painting Colors

Introduction

C hinese painting, with its long and glorious history, its several thousand years of intensive study, and the creativity of its countless artists, has developed its own unique form and style. Because the experience handed down to us from the past is so rich and extensive, nothing could be of greater aid to the continued development of our national painting tradition and to the creation of new works imbued with the realism of our national style than for us to apply a scientific attitude, combined with analysis, criticism, and receptivity. In inheriting our national painting tradition, we should not minimize the importance of understanding the special characteristics of materials and techniques, so how we go about understanding and using colors as they were once used in Chinese painting is an important issue. Colors are part of the material basis of Chinese painting and have an important function in the creation of a national style of art. While there is currently no shortage of artists, these artists are still not entirely familiar with the painting colors used in our magnificent national tradition. This situation, to a certain extent, can interfere with our work in "weeding through the old to bring forth the new, and letting a hundred flowers bloom."

Today we can still see quite a few ancient paintings done in color, important examples like the Han tomb murals at Liaoyang, Wangdu, and other sites, cave murals at various sites in the northwest, as well as paintings like *Traveling in Springtime* by the Sui Dynasty artist Zhan Ziqian (Palace Museum collection, Beijing), *Court Ladies Wearing Flowers* by Zhou Fang of the Tang Dynasty (Dongbei Museum Collection [now the Liaoning Provincial Museum in Shenyang]), *The Night Entertainment of Han Xizai* by the Five Dynasties artist Gu Hongzhong (Palace Museum Collection), *Lotus and Golden Pheasant* by Zhao Ji [Emperor Huizong] of the Song Dynasty (Palace Museum Collection), and other works. Although the oldest of these works are now about two thousand years old and the more recent ones eight or nine hundred years old, the gorgeous color of these paintings is still preserved, the various colors still retain an air of harmony, and they demonstrate the ancient masters' talent in the use of color.

In the literature on painting, more than 1400 years ago the artist Xie He propounded the "Six Principles," among them, "Conformity to kind in applying colors."[A] More than 1100 years ago, the art critic Zhang Yanyuan, in his *Record of the Famous Painters of the Successive Dynasties*,[B] enumerated the production sites for painting pigments, the nature of their use at that time, and the reasons for there being no change over time [in their use]. In each subsequent dynasty there were many artists who wrote, in varying degrees, about their experience with the use of colors. These documents are of great value in the study of Chinese painting colors.

Mineral Pigments

1. Reds

Cinnabar.[1] Also called Chensha cinnabar [from the name of a production site, Chenzhou, in Hunan Province], it occurs together with the Schreibersite class of ore.[2] The principle component is mercuric sulphide (molecular formula: HgS).[3] It is formed within limestone, in lumps like pillars and slabs, horses' teeth and arrowheads. China's important production sites include: Fenghuang, Huangxian, Mayang, and Gancheng in Hunan Province; Yuping, Bijie, Guizhu, and Anshun in

---

[1]Both in Chinese and in English literature, the distinction between cinnabar and vermilion is often not clearly maintained. As consistently as possible, we have used "cinnabar" to refer to the native mineral; "vermilion" (discussed subsequently) to indicate the synthesized chemical equivalent. Where the context itself offered no guide, we have translated *zhusha* 朱砂 and related terms as cinnabar; *yinzhu* 銀朱 as vermilion. This usage is affirmed by Song Yingxing's *Tian gong kai wu* of 1637 (Beijing: Zhonghua shuju, 1959), ch. 16, *xia*, p. 42a. Still, this terminology is not universally agreed upon, and some literature translates only *dansha* 丹砂 as cinnabar, while *zhusha* is translated as vermilion and *yinzhu* as deep vermilion—for example, Mai-mai Sze's translation of Wang Gai, *The Mustard Seed Garden Manual of Painting* (Princeton: Princeton University Press, 1956). In general, and in this case, we have followed the terminology used in Bernard E. Read and C. Pak, *A Compendium of Minerals and Stones Used in Chinese Medicine from the Pen Ts'ao Kang Mu* [*Bencao gangmu*] (Beijing: Peking Natural History Bulletin, 1928) and Bernard E. Read, *Chinese Medicinal Plants from the Pen Ts'ao Kang Mu, A.D. 1596: A Botanical, Chemical, and Pharmacological Reference List,* 3rd ed. (Beijing: Peking Society of Natural History Bulletin, 1936).

For additional details on the technology of cinnabar production, see Song Yingxing's *Tian gong kai wu*, ch. 16, *xia*, pp. 40a–43b, translated by E-tu Zen Sun and Shiou-chuan Sun (University Park: Pennylvania State University Press, 1966), pp. 279–85. For modern technical information and illustrations, see Rutherford J. Gettens, Robert L. Feller, and W. T. Chase, "Vermilion and Cinnabar," *Studies in Conservation,* 17, no. 2 (May 1972), pp. 45–69, where cinnabar and vermilion synthesized by sublimation are described as "entirely similar" even under a microscope and undifferentiable except possibly on the basis of impurities (p. 52). It is written there (p. 53) that vermilion "may darken in an apparently capricious manner." Peter Glum speculated that while cinnabar seems to be more stable than red lead and lead white, "in a water-soluble medium on the walls of caves cut into the living rock it might react to seeping and efflorescence even in a very dry climate. It also turns black under direct sunlight; what a thousand years of indirect sunlight might do we don't know"; Glum, "Meditations on a Black Sun: Speculations on Illusionist Tendencies in T'ang Painting Based on Chemical Changes in Pigments," *Artibus Asiae,* 37, nos. 1–2 (1975), p. 54. Chinese techniques to stabilize cinnabar and vermilion pigments by washing over layers of carmine are discussed on page 66, below. Like mineral pigments such as realgar, orpiment, azurite, and malachite, cinnabar produces a somewhat granular pigment with a slight sparkle, due to its crystalline structure and varying according to how finely the material has been ground. It yields a fairly opaque surface and coverage, but less so than azurite and malachite.

[2]Actually, the geological occurrence of cinnabar is by no means regular or predictable, although it often occurs in conjunction with hotsprings, volcanic or bituminous materials. Schreibersite (also known as Rhabdite and lamprite) has been variously and vaguely identified as a combustion product or meteoritic in origin, as a sulphide of chromium or an iron-based compound.

[3]This and several subsequent formulas have been altered by the translators from inaccurate ones that appear in the original text.

Guizhou Province; Xiyang, Xiushan, and Pengshui in Sichuan Province; and Baoshan and Dali in Yunnan Province; as well as other sites. The best is natural cinnabar that has a brilliant appearance, like a mirror. That which uses a refined mercury is not suitable for producing painting pigments.

Red standard. When cinnabar has been ground fine and a clear sizing solution is added, the material floating on the top, which is somewhat redder than red lead, is called red standard.[4] ("Standard" 標 is also written 膘 meaning "fat"—this being the part of the cinnabar that looks like grease floating on top, while the "standard" indicates that portion which has floated up, so the meanings are interrelated.)[5]

Vermilion. Also called purple acetate, this was the earliest invented chemical color in Chinese antiquity. In the past, the method of manufacture was as follows. Take one-half kilogram of mercury and grind it together with one kilogram of red sulphur [literally, "stone pavilion rouge"] (this is an herbal name; it is much like manufactured sulphur). Put it into a large-mouthed clay jar and cover the top with an iron pan, using wire to bind together the pan and the jar and daubing them with salt to seal them tight. Put this on an iron rack, with a charcoal fire below to heat the jar. While it is on the fire and for as long as it is being heated, use a palm brush dipped into cold water to brush cold water over the surface of the iron pan cover. It will be done in approximately one hour. After it has cooled off, lift off the iron pan and the sides of the pan and the jar will be completely covered with vermilion, while the red sulphur will remain sunken at the bottom of the jar. One-half kilogram of mercury is used to produce 700 grams of vermilion. After the Opium War, sea transport was opened up and mercury was exported, and since it was already expensive, this kind of refined vermilion was very little used. The most important production site in China was Zhangzhou in Fujian Province. Today, even so much as one packet of vermilion (weighing 50 grams) has been very difficult to obtain. Nowadays, mercuric sulphide is used as a substitute.[6]

Red ochre. Also called earth red, some of this pigment is made through heat processing, some through processing in water. The red ochre that goes into painting is extracted from red iron ore [ferric oxide, molecular formula: Fe_2O_3].[7] A good raw material is that which is extracted from red iron ore and rubbed by hand until it feels creamy in texture. The raw product comes from the Yanmen region in Shanxi Province, which in the past was known as Dai prefecture, so that this pigment is also called "Dai ochre." In all the places where red iron ore is found, red ochre is produced.

---

[4]Red standard is a yellowish red pigment but redder than red lead. It is also called Satsuma red. The character *biao*, translated here as standard also signifies a (high) mark, a flag, the twigs of a tree, an edge, or a limit, all of which relate to the notion of a material which floats toward the surface when ground up in solution. Other mineral pigments such as vermilion, azurite, malachite, and yellow ochre produce a similar "standard" in solution (see chapter 5, section 2, for further details).

[5]When ground cinnabar is placed in a liquid suspension, the material known as "yellow fat" floats to the surface as a thin layer just above the red standard. The yellow fat is discarded and not used in painting.

[6]Red ochre (ferric oxide) and yellow ochre (ferric hydroxide) may each be produced from the other, red ochre made into yellow ochre by hydrating, yellow ochre into red ochre by heating; this is apparently what is referred to by the author as heat and water processing. Compared to cinnabar, red ochre is a darker, somewhat brownish color, and much more opaque. Red and yellow ochres, which were among the first pigments used by early man in China, are virtually permanent.

[7]See Kazuo Yamasaki, "Chemical Studies on the Eighth-Century Red Lead Preserved in the Shōsō-in at Nara," *Studies in Conservation*, 4, no. 1 (February 1959), pp. 1–5.

Red lead [or minium, sometimes referred to as lead yellow, a lead oxide, molecular formula: Pb_3O_4]. Also called Zhang red, this is a product of Zhangzhou, in Fujian Province. Nowadays, every shop that sells painting pigments stocks it. The method of manufacture consists in using the lead that is left over from making lead white, which, when heated up again, produces red lead.[8]

2. Yellows

Mineral yellow, realgar, orpiment, and yellow ochre are differentiated according to the depth of their color, since they all occur together. Consisting of arsenic trisulphide (molecular formula: As_2S_3), mineral yellow is a true yellow color; realgar [also called cockscomb yellow or red orpiment, arsenic disulphide, molecular formula: As_4S_4] is an orange color; orpiment is a golden color; and yellow ochre [or earth yellow, ferric hydroxide, $Fe_2O_3 \cdot H_2O$] is the yellow of loess soil.[9] Gansu is the principal production site. Of special note is the fact that Hunan Province has the world's largest realgar mineral deposits. These four pigments should not be used together with lead-based pigments.[10]

Mineral yellow. Also called gold-stone [which is usually identified as realgar], its surface is coarse and its color is dark, and it has an unpleasant smell. The first or second layers beneath the surface provide a good mineral yellow.

Realgar. This is found either within the mineral yellow, covered over by it, or in lump form without any kind of covering. There is also some realgar that is suffused with a glossy sheen, with a much deeper color, that is called essence of realgar.

Orpiment. Orpiment is also found within mineral yellow. It occurs in layers resembling mica. As it is very easily fragmented, there is a folk adage that says, "From just four taels of orpiment come a thousand layers of gold leaf."[11]

[8]This sentence is a bit confusing, since vermilion itself is mercuric sulphide; perhaps the author means native cinnabar. In Chinese, red lead is literally "yellow cinnabar," as it is more yellow or orange than cinnabar, varying quite a bit in its yellow content. It is very opaque. Like other lead-based pigments, red lead is not stable when exposed to light but turns brown or black, as it has at Dunhuang. Moreover, in combination with lead white it turns brown, as so often happened at Dunhuang when lead white was used to highlight red lead-based skin; cf. R. J. Gettens' chemical analysis in Langdon Warner, *Buddhist Wall Paintings, A Study of a Ninth Century Grotto at Wan Fo Hsia* (Cambridge: Harvard University Press, 1938), pp. 9–10, and Glum, "Meditations on a Black Sun," pp. 54–55. In combination with sulphide-based pigments, lead red turns black. See also Song Yingxing, ch. 14, *xia*, pp. 23a–23b, translated by Sun and Sun, p. 256 (lead red, *huangdan*, is rendered there as "litharge," although litharge is lead monoxide, PbO); Song's chapter 14 is devoted to the production of metals, including lead, iron, copper, gold, and silver.

[9]A clarification is required here: mineral yellow is often used as a synonym for both realgar and orpiment or a broader term to include both; although the author treats mineral yellow as a distinct pigment, the chemical identification given here, arsenic trisulphide, is that of orpiment, while a synonym given for it here, literally "gold-stone," is identifiable as realgar. Also, yellow ochre (ferric hydroxide) differs chemically from the others, although it possibly occurs together with them. See Read and Pak, *A Compendium of Minerals and Stones*, pp. 36–38.

[10]Orpiment, or mineral yellow, and realgar may darken the color of lead white; yellow ochre, on the other hand, can be used safely with lead-based pigments. Realgar and orpiment also should not be combined with cinnabar because of its mercury base, nor with the copper-based pigments, azurite and malachite, nor with the iron oxides, yellow ochre and red ochre; cf. Jean Ippolito, "Chinese Paint Pigments and Their Classification, with an Examination of Four Ming Dynasty Paintings at the Seattle Art Museum" (M. A. thesis, University of Washington, 1985), p. 40. Realgar and orpiment may also darken within a few months' exposure to light; Ippolito, "Chinese Paint Pigments," pp. 40–41.

[11]Orpiment, an alteration of realgar, is usually found as a layer surrounding a realgar core.

Yellow ochre. This is the deepest yellow color, unpleasant-smelling, enclosed within the outer portion of the mineral yellow. Actually, all of the pigments mentioned above are somewhat foul-smelling.[12] Its principal components are ferric oxide and ferric hydroxide. The excess is used as potter's clay.

3. Blues

In his *Supplementary Notes on Famed Physicians,* Tao Hongjing of the Liang Dynasty (a painter and also a student of herbal medicines, A.D. 452–536) wrote, "Azurite occurs where copper is found." This saying accords with modern scholars' recognition that azurite is a basic carbonate of copper—molecular formula: $Cu_3(OH)_2(CO_3)_2$—extracted from red copper ore. Its five varieties are azurite, flat blue, layered blue, light blue, and granulated blue, all of them [slightly] toxic. [The latter four are treated here as varieties of azurite, differing primarily by virtue of their particulate structure and range of hues.] They may be discussed separately as follows.[13]

Azurite. Lumps of this material resemble the red bayberry. According to what Su Song of the Song Dynasty (ca. 1058) said, "Recently, there has been a type from Raozhou and Xinzhou [both in Jiangxi Province] whose form resembles the red bayberry, having a hollow core and when crushed yields a thick liquid, but it is hard to obtain." (See the *Illustrated Classic of Medicinal Plants* [Li Shizhen's *Bencao gangmu* of 1596].) Herbalists and artists prior to the Northern Song Dynasty readily discussed azurite, saying that it came from gold ore or copper ore. I have only seen azurite resembling the fruit of the red bayberry produced in Sichuan Province, but in the center there were only a few small hollows and no thick liquid, and it was not very usable.

Flat blue. Also called "pure blue," it is found in both Yunnan Province in China, and in Burma. That which is made in Yunnan Province is called Yunnan blue (Dian 滇 *qing*), and that produced in Burma is called Burmese blue (Dian 甸 *qing*). This is the azurite which the Qing Dynasty author Wang Gai calls plum petal (see *The Mustard Seed Garden Painting Manual*). While the pieces of pigment produced in Burma are larger, it is not as delicate and beautiful a color as Yunnan blue. [Flat blue is regarded as the highest quality azurite pigment.]

Layered blue. The character for "layered" 曾 should be written 層, as in 層次 "sequence." This has either one dark and one light layer, or it has several layers all

---

[12]These arsenic-based compounds give off dangerously toxic gases when exposed to sunlight and are now seldom used, even banned in many places.

[13]Technical information on azurite may be found in Rutherford J. Gettens and Elizabeth W. Fitzhugh, "Azurite and Blue Verditer," *Studies in Conservation,* 11, no. 3 (August 1966), pp. 54–61. Read and Pak, *A Compendium of Minerals and Stones,* p. 57, and some other sources treat the two terms used here for azurite (*kong qing*) and layered blue (*ceng qing*) as forms, instead, of malachite (as the hollow variety and the stratified variety, respectively). Their identification is inconsistent with the author's, as well as with the blue, not green, colors which *kong qing* and *ceng qing* primarily produce. Because of their close occurrence in nature, malachite and azurite rarely occur without impurities of the other mixed in. Azurite often comes in the form of small, round, marble-sized nuggets, sometimes hollow (*kong*) and sometimes with green malachite in the center (for verification of the latter, see Ippolito, "Chinese Paint Pigments," p. 9). This joint occurrence permits some acceptable variation in classification and translation. Azurite produces a slightly granulated pigment that can yield a quite opaque coverage of the painted surface. It normally remains stable when exposed to sunlight and normal atmosphere.

of which are a deep shade of blue. Artists are fond of its lighter blue color, and gather together a quantity of this light color to refine at one time. The light blue that comes out of the refining process is called "sky blue." It comes from Shanxi, Hunan, Sichuan, and Xikang provinces and from Tibet [Xikang is now divided between Sichuan Province and Tibet].

Light blue. Also called jade blue, this comes from the provinces of Yunnan, Guizhou, and Sichuan. It has a lighter color than "sky blue" and lacks its luster, and artists seldom use it.

Granulated blue. Also called Buddha blue [the traditional color of the Buddha's hair] and Hui blue [meaning Moslem blue, in reference to the region from which it was imported]. This is a pigment that came to China from the Western Regions. Although among China's ancient texts there are no clear records of it, in Buddhist paintings and in colored architectural renderings on silk, in the Dunhuang murals and Buddhist images of the Ming and Qing dynasties, granulated blue was always used. It is divided into two types, like coarse and fine sand; the coarser type has small pellets the size of grains of millet, while the finer type has smaller particles although it is not a powder.[14] Packages of granulated blue weigh 2,400 grams. It is still found today in Xikang and Xinjiang Provinces and in Tibet as well as in other sites. Folk artisans call the pigment produced in Tibet "Tibetan blue" [a deep blue color].

4. Greens

Fan Chengda of the Song Dynasty, in his *Gui hai yu heng zhi,* wrote, "Mineral green comes from along the You River [in Jiangxi Province] where copper is located, found embedded in rock. This sort of mineral is called malachite. There is also a type fragmented like clods of earth that is called paste green." This, and what today is referred to as malachite [$CuCO_3 \cdot Cu(OH)_2$] extracted from copper ore, are the same. It is toxic. The division into four varieties is as follows.[15]

[14]Granulated blue is equated here with the imported blue used since early times for the hair of the Buddha. Ordinarily, the blue used for this was ultramarine (derived from lapis lazuli, a sulphur-containing sodium aluminum silicate, approximate formula $Na_{8-10}Al_6Si_6O_{24}S_{2-4}$), which was probably exported to China from the Badakshan mines in northeastern Afghanistan. A second alternative was smalt, prepared from a ground glass (potassium-based, silicate, but varying considerably in specific chemical composition) whose blue color was produced by small amounts of cobalt oxide added during manufacture; smalt was also Near Eastern in origin and appeared in Chinese paintings as early as the eleventh century. The fine and coarse varieties of granulated blue seem to correspond, respectively, to ultramarine and smalt (both being silicates). However, Yu Feian firmly suggests (both here and below, p. 71) the identification of granulated blue as a form of azurite. See Joyce Plesters, "Ultramarine Blue, Natural and Artificial," *Studies in Conservation*, 11, no. 2 (May 1966), pp. 62–91, and Bruno Mühlethaler and Jean Thissen, "Smalt," *Studies in Conservation,* 14, no. 2 (May 1969), pp. 47–61.

[15]Technical information on malachite, with illustrations, may be found in Rutherford J. Gettens and Elizabeth W. Fitzhugh, "Malachite and Green Verditer," *Studies in Conservation*, 14, no. 1 (February 1974), pp. 2–23, where it is noted that "a special characteristic of massive malachite is color banding ranging from near black to pale green" (p. 2). As noted above, malachite and azurite tend to occur interspersed. Like azurite, malachite produces a pigment that appears slightly granular on very close inspection and which is fairly opaque. Like azurite, it is regarded as stable in light and in normal atmospheric conditions; Gettens notes that it is "theoretically subject to blackening when mixed with sulphide pigments" [such as orpiment or realgar] but that "in practice darkening from this cause apparently has not been reported" (p.7).

Malachite. This is found in lumps. The best malachite in lump form comes from Huize, Dongchuan, and Gongshan in Yunnan Province, while that from Nandan and Pinyang in Guangxi Province is somewhat inferior. There is also a large-lump form of malachite produced in Persia and Burma.

Peacock green. This also occurs in lump form. It naturally occurs with a mixed pattern of dark and paler tones, strongly resembling the emerald green color of a peacock's plume [bluish-green or turquoise in color.] It comes from the northwest of China and from the Malay Peninsula. It is used in folk handicrafts for carved and inlaid ornaments. Even when broken up into bits and pieces, a green color can still be produced from it.

Verdigris. Also called copper green. Resistance to sunlight is its outstanding feature. This also, in its natural state, occurs in copper ore.[16] The ancient method of manufacture is this: "Strike off slices of copper, soak them overnight in good vinegar, wrap them in husks, smoke them over a low fire, then scrape off the verdigris that has accumulated." This is also one of China's earliest invented chemical pigments. (See Li Shizhen's *Compendium of Medicinal Plants,* published in 1590 [1596], and Song Yingxing's *Tian gong kai wu,* or *On Using the Products of Nature,* published in 1637.) [This process produces a copper acetate, not highly regarded as a painting pigment.] The modern method of manufacture is to add sodium carbonate to a solution of blue vitriol (chalcanthite) to obtain the precipitate. [This produces a basic copper carbonate, a synthetic equivalent of malachite.][17]

Granulated green. This comes from Tibet and Persia. Granulated in form, the color of this pigment is rather deep.

5. Whites

In Chinese painting, the white pigments are in the same category as the coarse blues and greens, that is, all are considered to be "thick colors." (This is in distinction to the lightly applied colors.) These include chalk, lead white, and clamshell white. Clamshell white is considered to be a mineral pigment because, although it is made from calcined shell, when it undergoes the process of calcination it is transformed into lime.

White chalk. Also called "white powder." It is composed of calcium carbonate (molecular formula: $CaCO_3$). This substance was extremely important in early Chinese painting. Before A.D. 536, it was known as "painting white" (according to Tao Hongjing of the Liang Dynasty). This was a principal pigment used in murals from the Han and Wei Dynasties onward. It is found nearly everywhere; the best of it is produced in Hebei, Shanxi, Anhui, and Henan provinces. It does not change in color over time. On the Northern Wei Dynasty murals at Dunhuang, it was blended with vermilion of Zhang red [red lead] to render the flesh tones of

---

[16]That this material occurs naturally seems doubtful.

[17]For technical and illustrated information on verdigris, see Hermann Kühn, "Verdigris and Copper Resinate," *Studies in Conservation,* 15, no. 1 (February 1970), pp. 12–36. Verdigris is described as a "collective term for copper acetates of different chemical composition, which range in colour from green, via green-blue and blue-green, to blue" (p. 12). Verdigris pigments are subject to a shift from blue-green to green, taking place particularly rapidly at first, primarily during the first months' time after preparation (p. 16). Several variant molecular formulas are noted, such as one given for a green-colored verdigris as $Cu(CH_3COO)_2 \cdot [Cu(OH)_2]_3 \cdot 2H_2O$.

faces, bodies, and hands, but since that time the vermilion and Zhang red have turned a blackish color, which of course affected the chalk.[18]

Lead white. Also called foreign white, Mandarin white, and zinc flower.[19] Its component is a basic lead carbonate, molecular formula: $2PbCO_3 \cdot Pb(OH)_2$. It was, in the past, the white pigment with which women made up their faces. Because it is manufactured in the shape of silver ingots, it is also known as "ingot white." It is another of China's ancient pigments produced through chemical methods. It is said to be a method that Zhang Qian of the Han Dynasty brought back when he was sent as an envoy to the Western Regions. The way it is made is this: take 50 kilograms of lead, smelt it into thin sheets, roll them into lead tubes, and place them in a wooden barrel. Then pour a bottle of vinegar into the bottom of the barrel and another one into the middle area. Cover the barrel tightly with a lid and seal it with paste and paper so that no air is allowed to escape. Fire it in a vent stove and after seven days open the lid and white crystals will have formed all over the lead and the barrel. Brush these crystals off into a jar, again place the lead tubes in the barrel, and continue the process in this manner until the lead is eventually used up. Each time white crystals are brushed off, add 100 grams of bean starch and 200 grams of clamshell white and this will make lead white. If the lead white pigment is placed in a charcoal stove, it will revert to lead. Therefore, when lead white is used in painting and with the passage of time turns a blackish color this is referred to as "reverted lead." However, if a light wash of hydrogen peroxide solution is applied it will recover its white color.[20]

Clamshell white [molecular formula: $CaCO_3$]. Also called "pearl white," this had an important use in early Chinese painting. It was always used in the Song Dynasty as a substitute for chalk. The method of preparation is: select ocean clams that have a thick, sturdy shell and a faint ring of purplish-red color around the mouth. Over a low fire, the shells become quicklime, which is ground extremely fine until it becomes a white powder. With the addition of water, this is transformed from quicklime (shell lime) into lime. When glue is added for use, it will not change color. [Rather, it becomes whiter and increasingly opaque with the passage of time.]

6. Black

Black paste. Produced in Hubei and Hunan provinces, black paste [or cosmetic

[18]See Rutherford J. Gettens, "Calcium Carbonate Whites," *Studies in Conservation*, 19, no. 3 (August 1974), pp. 157–84.

[19]Yu Feian regards this "zinc flower" as a lead white. More likely, it is zinc white (zinc oxide, ZnO), a white powder obtainable, among other means, by heating zinc ore together with charcoal in a reduction furnace. The author discusses zinc white elsewhere. In more recent times, zinc white and titanium white (Ti_2) have largely replaced other white pigments, due to their greater whiteness, opacity, and stability.

[20]A more complex method for restoring lead white to its original color is given below, p. 62n. Lead white (unlike chalk and clamshell white) also darkens on contact with sulphides. The virtue of lead white, balancing its lack of durability, is its greater opacity and covering capacity than calcium-based white pigments. Lead white was the most expensive of these white pigments. For further technical information, see Song Yingxing, ch. 14, *xia*, 22b–23a, translated by Sun and Sun (as "ceruse"), p. 256; Rutherford J. Gettens, Hermann Kühn, and W. T. Chase, "Lead White," *Studies in Conservation*, 12, no. 4 (November 1967), pp. 125–39; John Winter, " 'Lead White' in Japanese Paintings," *Studies in Conservation*, 26, no. 3 (November 1981), pp. 89–101. As reported in Winter, one "unusual" occurrence of lead white in a scientifically tested Chinese painting proved to be not a lead carbonate but a lead sulphate.

black] is used as medicine and can be purchased at Chinese pharmacies. It is also known as graphite. It has the quality, when placed in the mouth, of sticking to the tongue, unlike coal. It is principally composed of carbon. The painters of the past ground it up fine and used it for painting beards and eyebrows.

The preceding are the mineral pigments most frequently used in modern times. There are other less generally used colors, such as agate, coral, gemstone, turquoise, and amber, about which the reader will forgive me for not going into detail.

Plant Pigments

1. Safflower

This plant [*Carthamus tinctorius*, a member of the thistle family] closely resembles the indigo plant, having globe-shaped clusters of blossoms. If the flowers are picked in the early morning, then in a day or so new blossoms will issue forth from the cluster, and this should continue until there are no more to pluck. The blossoms are ground to pieces and wrapped in cloth and a yellow juice extracted from them, then they are placed in the shade to dry and molded into cakes. When it is time to use them, steep them in warm water until they are fully suffused. Then [adding a mild alkali such as sodium carbonate or the extract of plant ash to draw the red dye from the plant material], squeeze this liquid out with a cloth and add glue for use. These days, only in the minority peoples' areas is this still used as a red dye. In the past, red paper which we used for celebrations was always dyed with safflower or madder. After the Opium War, pigments like carmine and magenta were imported in great quantity so that safflower and indigo were gradually replaced in the pigment market by imported products.[21]

2. Madder

This is a creeping plant [*Rubia tinctorum*] with leaves that resemble the Chinese date and square stems that are hollow in the center. Each segment puts forth five leaves which bloom with red flowers. The crimson roots are boiled to extract a liquid that produces a red color. Nowadays, madder still grows wild in the provinces of Hebei and Henan and in the northwest. The color of madder is redder than that of safflower.[22]

---

[21]Safflower pigment is described as "yellowish red" by Rokurō Uyemura, "Studies on the Ancient Pigments in Japan," *Eastern Art* 3 (1931), p. 57. Like many plant pigments and perhaps more so than most, safflower pigment is translucent and weak in its covering capacity, varies considerably from specimen to specimen, and tends to be fugitive on exposure to light. For brief information on cultivation of the safflower plant and its use as the basic red dye for cloth, see Song Yingxing, ch. 3, *xia*, pp. 51a–52a, translated by Sun and Sun, pp. 76–77.

[22]The primary coloring ingredients in madder are alizarin, $C_{14}H_6O_2(OH)_2$, and purpurin, $C_{14}H_5O_2(OH)_3$. Joseph Needham has reported it as remaining relatively stable when exposed to light, *Science and Civilization in China*, Vol. 5, part 2 (London: Cambridge University Press, 1974), p. 29.

3. Lac

[Literally, "purple ore."] Also called "purple stem," and "purple shoots," this is produced in the southwest border regions of China. It is used in medicine and it is also essential to the electrical industry. Lac is a kind of natural tree resin. As long ago as the Tang Dynasty, Zhang Yanyuan referred to it as "ant-ore," which is a material that makes a purplish-red colored pigment.[23] Lac does not dissolve in water, but when ground up fine and mixed with glue it can be used.

4. Rouge

Rouge can be made from the previously discussed materials: safflower, madder, and lac. It is related that "in 1183 B.C., during the reign of King Zhou of the Shang Dynasty, the people used the juice of the safflower to create rouge, which was applied by women as 'peach blossom makeup.'" (See *Zhonghua gu jin zhu.*) It is also said that when Zhang Qian of the Han Dynasty was an envoy to the Western Regions, he returned from the country of Yanqi with Yan *zhi,* that is to say, rouge. The cakes of rouge that are used as women's makeup are made in the same ingot shape as the cakes of lead white makeup. Most recently, "foreign rouge," that is, packaged cosmetics, have replaced it. Until now, genuine rouge has always been quite difficult to obtain. The cakes of rouge that I have previously obtained from Guangdong Province are of a color close to purple and are made from lac. The rouge cakes I have found from Fujian Province and the cotton-ball rouge from the city of Hangzhou both are manufactured from safflower and madder. I have heard that the rouge from Gansu and Xinjiang provinces, and from the southwestern border and other areas, is an especially intense red, but I do not know if any of it exists any longer. Rouge was always the principal color used in ancient times for

---

[23]Contrary to this view, which is a commonly held one, it has been shown that lac (or gum-lac) is not a natural tree resin generated in defense against wood-boring insects but rather is a product secreted by certain insects for their own self-protection. In early times, for the Chinese, this insect was the *Laccifer lacca,* also known as *Coccus lacca,* a native of South Asia. In later times, lac was largely replaced by "Western red" or carmine, derived from a relative of this insect, the *Dactylopius coccus* or cochineal insect, a native of the Americas. Cochineal dye, from which carmine is produced, was first brought to Europe after 1518 by Spaniards and subsequently transported by them to the Philippines and the rest of Asia. The term "ant-ore" used by Zhang Yanyuan in the ninth century probably reflects his early awareness of the source of this pigment as being an insect. In his discussion of carmine, on p. 70, below, the author states that the pigment is based on an animal substance. The *Laccifer lacca* females produce both a resin (lac) and a mixture of red anthraquinone compounds (lac dye). The crude product incorporating both of these still coating the twigs of trees, which are broken off and dried to kill the insects inside, is called stick lac; this is probably what is referred to as "purple shoots" and "purple stem" lac. Granulation of stick lack and extraction of most of the dye components with water leaves seed lac; further purification provides the shellac used industrially. Lac should not be confused with lacquer, a colorless fluid derived from the natural resins of the lac tree (*Rhus verniciflua* according to most botanists, *Rhus vernicifera* according to some of the practical literature). Aside from its use as a base for pigments in the painting of lacquer utensils and furniture, which flourished greatly in China in the Warring States through early Han periods, lacquer was only rarely used in later Chinese painting, as, for example, in building three-dimensionally the pupils of birds' eyes in Zhao Ji's (Song Emperor Huizong's) *Finches and Bamboo.*

See Berthold Laufer, *Sino-Iranica* (Chicago: Field Museum of Natural History, 1919), pp. 476–78; Edward Schafer, "Rosewood, Dragon's Blood, and Lac," *Journal of the American Oriental Society,* 77, no. 2 (1957), p. 135. Ippolito, "Chinese Paint Pigments," p. 97, reports about variability that a sample of newly applied lac pigment changed to a deep, dark red in a year's time, while a similar sample of applied cochineal pigment did not change significantly. Lac pigment can be expected to become duller and less saturated in time.

the crimsons and purples in Chinese painting and for the red washes that were used over cinnabar. However, with the passage of time, paintings done with rouge have faded in color. Today, "Western red" or carmine is used as a substitute and is more brightly colored.

5. Sandalwood

This is also known as logwood or sappanwood and is used to dye wooden objects. Its color is a deep purple, and it can be boiled and condensed into a paste for use.

6. Rattan yellow or gamboge

Rattan is the garcinia [*Garcinia hamburyi*], a deciduous tree that reaches a height of fifty to sixty feet. It is a tropical plant of the Guttiferaceae family. When holes are bored in the bark of the tree, its sticky yellow sap is allowed to flow out and is caught in bamboo tubes. After it has fully dried out [in the tubes] it is slightly hollow down the middle, and you have what is used in Chinese painting as "brush-handle rattan yellow." Like azurite, malachite, and verdigris, mentioned in the previous section, rattan yellow is toxic and should not be put in the mouth.[24]

When rattan yellow is sold at Chinese pigment shops, it is always called "moon yellow," *yue* yellow, because the finest is produced in Yuenan [Vietnam]. The next best in quality comes from Burma and Thailand. Shopkeepers have simplified the writing of *yue* for "Viet" 越 to *yue* for "moon" 月 so that to this day it is nicknamed "moon yellow." This pigment was already imported to China before the Tang Dynasty and was known as "Zhenla painting yellow" and "Linyi yellow."[25]

7. Sophora yellow

When the pistils and stamens of unopened flowers of the *Sophora japonica* [the locust or Chinese scholar-tree, called *huai* in Chinese] are used, a light yellow-green color is produced. Working with flowers that have already opened yields a yellow-green color. The method of manufacture for both is to scald the flowers in boiling water, then to mold them into cakes; these can then be used by squeezing the liquid out of them [after they have been remoistened]. When using malachite especially, this preparation must be employed as a covering wash [a layer of Sophora yellow being applied over malachite in order to improve its adherence to the surface].[26]

8. Cork tree bark

Called "yellow-wood" in Beijing, it comes from Sichuan Province. The color is a

---

[24]This refers to the painter's technique of bringing the brush tip to a point with the lips, which should not be done when using certain pigments. Rattan yellow is a pure, extremely bright yellow.

[25]In Tang times, Zhenla was a small state located in the northern portion of present-day Laos and Vietnam. Linyi was a small state on the Malay Peninsula.

[26]Song Yingxing noted that the immature flowers were the essential ingredient for green cloth-dyeing in China, and that the trees do not bloom before their tenth year (ch. 3, *xia*, pp. 52a–52b, translated by Sun and Sun, p. 77).

deep yellow and it can be used to repel moths. It can be decocted into a liquid and, with the addition of glue, condensed into a paste for use.

9. Gardenia seed

These can be bought at Chinese pharmacies. They are pounded to remove the skins, then boiled, and glue is added for use. This can serve as a substitute for rattan yellow.

10. Indigo ["flower blue"]

This is manufactured from the indigo plant. Indigo is used as a dye in China, having been discovered at a very early date. Ancient works such as the "Yue ling" ["Monthly Orders," from the *Li ji*] and the *Shuo wen jie zi* all spoke of the indigo plant as one which yields a blue dye. By the last years of the Guangxu era [1875–1908], cloth-dyeing in every locale had gradually switched over to the use of foreign indigo, while Chinese painters employed Prussian blue in place of indigo. Nowadays, the indigo plant is grown only by the Miao people in the southwest of China, who still use it to dye fabric. Indigo is a more brilliant color than Prussian blue and is able to tolerate sunlight without changing color too much.[27]

Indigo is a plant of the Polygonum or knotweed family [including *Polygonum tinctorium*, genus *Indigofera*; a number of related dye-producing plants were grouped together by the Chinese as the indigo plant or "blue plant."] In its first year, it puts out roots and the stems grow to a height of two or three feet. The leaves have an elliptical shape and at the base of the leafstalks there are tubular protective stipules that surround the stems. In autumn, the axils put out long stems which bear spikes of small red flowers at their tips, with red-colored calyxes. The leaves are the raw material from which indigo pigment is made. There are four or five varieties, all of which can produce indigo.[28]

The method for making indigo is this: the leaves of the indigo plants are gathered in autumn and spread out and piled in layers on a wooden board. Water is sprinkled on them and they are covered with a hemp bag until they begin to generate heat and ferment. After they have fermented, the hemp bag is removed and they are allowed to dry. The upper and lower layers are mixed, again sprinkled with water, and again allowed to ferment. The process is repeated

[27]Although it is regarded as fairly stable, some evidence of long-term deterioration exists. At the Byōdō-in at Uji, where an indigo was mixed with yellow ochre as a substitute for azurite, "the original blue color was almost completely lost due to exposure in the air for about 900 years, but from the margin of the painting concealed under a frame a bluish color appeared during the repair work, and indigo was chemically identified"; Kazuo Yamasaki and Yoshimochi Emoto, "Pigments Used on Japanese Paintings From the Protohistoric Period Through the 17th Century," *Ars Orientalis,* 11 (1979), p. 9. Ippolito, "Chinese Paint Pigments," p. 69, adds that the stability of indigo depends also on the mordant used.

[28]Song Yingxing said there are five kinds of indigo plant that yield indigo dye, including *Polygonum tinctorium*, "horse indigo" or *strobilanthes flaccidifolius*, "Kiangsu indigo" or *Indigofera Kiangsu*, "tea indigo" or *Isatis tinctoria*, and a small-leafed variety of *Polygonum tinctorium* commonly called "pigweed indigo" or *Amarantaceae tinctorium*; ch. 3, *xia*, p. 50a ff., translated by Sun and Sun, p. 75 ff.

several times, until there is no more fermentation, and thus natural indigo is produced. This is a cleaner method than steeping them in lime.[29]

Painters take the indigo made in this way and grind it with a mortar and pestle. It requires eight hours to grind up approximately 200 grams of indigo. After it has been ground, liquid glue is added and it is left to settle. After it has settled, the material that has floated up to the top is skimmed off, which is the good indigo [literally, "flower blue," the lightest colored indigo] that we want.[30]

11. Ink

This provides the primary color of Chinese painting—black. Ink came into existence through the creativity of the working people of China and it is famous throughout the world. There are three types of ink: pine-soot ink, oil-soot ink, and lacquer-soot ink, all made principally in Huizhou [in Anhui Province].

12. Vegetable-soot

This is also called "chimney ink" and "pan-bottom soot." This is the black soot obtained from burning wood or vegetable matter, with glue added for use. In painting beards and hair, as well as for feathers and fur, this is always used.

13. Rice-paper plant ash

This is also called "rush candle ash." The stem pith of the rice-paper plant [*Tetrapanax papyferus*], which is used in Chinese medicine, is put into an iron tube where it is burned to produce ash, then glue is added for use. This is the black color that is used exclusively for painting moths and butterflies.

The above are the commonly used plant pigments, which Chinese painters generally call "vegetable colors." They are named in distinction to the mineral pigments, which are generally known as "mineral colors."

Carmine [also "Western red" or cochineal] was adapted for use at an early time by Chinese painters. This pigment is a precipitate made from animal material [the cochineal insect].[31] It does not soak through to the back of the paper or stain the hairs of the brush. In pre-war Germany, it was produced as a powder by the Greater German Pigment Company and in cake form by the Pelikan Company, both with glue already added. That sold by the Jiang Sixu Hall of Suzhou is of the former type.[32] Whether one adds warm or cold water to these two types of pigment, neither has a foul smell; both are quite expensive. Another type of carmine is made in England, and it is rather dark and deep in color. It, too, does not stain the brush hairs or soak through to the back of the paper, but it has a foul smell.

---

[29]The primary dyeing component of indigo is indigotin, molecular structure: $C_{16}H_{10}N_2O_2$. This is derived through the fermentation of a glucoside contained in the indigo plant.

[30]Before it has settled into different layers, indigo pigment is a dark color that resembles ink.

[31]Like lac pigment, carmine pigment produces an intense, purplish-red pigment. See footnote 23 on lac and cochineal, above.

[32]This famous shop is still the major source in China for widely distributed pigments of fine quality. For a brief note on the history of the Jiang Sixu Hall, see below, p. 30.

Gold and Silver

Although gold and silver cannot be mixed with other colors to produce secondary colors, they cannot be slighted with regard to Chinese painting, and so they are added to our discussion here.

1. Gold

All the gold that is used in Chinese painting is produced from gold that has been hammered into foil. This includes gold paste, sprinkled gold, and gold leaf. Gold foil is a specialty of Suzhou, of which there are two types: "pure gold" and "Buddha pure gold" [the latter, a pure gold with a reddish cast]. "Pure gold" has the basic color of gold, while the latter is purer still. There is also "*tian* pure gold," which is a light yellow color. Each of these takes ten sheets of foil to make one "book" and a thousand "books" to make a "box." Sprinkled gold is not used in Chinese painting. "Gold ground" is applied only to paper or silk before a composition is completed, laying the gold as a foundation in the areas that have been left empty, after which one may again proceed to add colors. "Gold ground" may be divided into "raindrop gold" [perhaps a sprayed gold], "fish-roe gold," "gold frost," and other types, and only Suzhou has specialists in this. "Gold paste" is made in a dish, adding glue and using the fingers to mix gold foil into a fine paste. Using a brush dipped in it, it can be applied in drawing. In Suzhou at the Jiang Sixu Hall Pigment Shop, they have pellets made of this paste as well as gold shaped as bowls and attached with glue to the surface of small porcelain cups. The Jiang Sixu Hall Pigment Shop also stocks foreign gold in paste form which is comparatively inexpensive.[33]

2. Silver

In painting, silver is rather rare and used only in horse saddles, knives, and spears. Silver foil is also produced in Suzhou. As with the method for making gold paste, it is finely mixed with the fingers and dipped into with a brush for use. There are some who prefer to grind up silver foil, mercury, and table salt finely, using a mortar and pestle, after which the mercury is driven off with heat (adding a bit of alcohol to heat it with) and the salt is rinsed off. This method saves time and makes the grinding easy.

Silver foil used together with realgar and smoked until it is a fine paste can be used for gold but the color will fade over time. If, after using this silver paste in painting, a wash of gardenia yellow is applied over it, the gold color will not change over time.[34]

In paintings, even though true gold and true silver are used, if they are lacking in luster people will always think that the gold and silver are fake. This is actually

---

[33]For information on gold production and use in China, see Song Yingxing, ch. 14, *xia*, pp. 1b–4a, translated by Sun and Sun, pp. 235–38.

[34]For more on silver production, see Song Yingxing, ch. 14, *xia*, pp. 4a–6b, translated by Sun and Sun, pp. 238–42.

due to faulty application. In using them, one must be sure to submerge them thoroughly in the glue, forcing them to sink to the bottom of the dish and using a brush that has been dipped to the bottom of the dish, for then they will take on a natural luster. Gold and silver are thus different from the other colors in the manner of their application.

Glue and Alum

That the colors of Chinese paintings have remained bright, lively, and enduring over a long span of time is the result of the ceaseless efforts and creativity of the painters of China. On the one hand, there is the selection of raw materials and the process of refining them, and on the other, the use of glue and alum to fix these materials so they will not disintegrate and flake off of the surface. Even though a pigment may not dissolve in water, still it is necessary that its brightness of color remain unchanged over time, as well as that it remain affixed to the surface of the painting—which is the function of glue and alum. Now these are set forth as follows.

1. Oxhide or "yellow-transparent" glue

This is also called "Guang glue," as it is produced in the provinces of Guangdong and Guangxi. It is manufactured from the hides, sinews, bones, and horns of oxen and horses. (This is a new substance formed from the hard protein of the hides, sinews, bones, and horns of oxen and other animals when it is broken down under the effects of water and heat.) Of a yellow hue and transparent, it is formed in rectangular strips and is not unpleasant smelling. It is melted in water over a low fire. Only the light, upper, clear layer is used to add to colors; the turbid material below is not used.

2. Donkeyhide or "E" glue

This is also called "preserved glue." It, too, is manufactured from the hides, sinews, bones, and horns of oxen, horses, and other animals. It comes from Ejing, thirty kilometers to the northeast of Yanggu in Shandong Province. There are three types of donkeyhide glue: the one that is clear, thin and transparent, and of light yellow hue is the choice for adding to pigments. There is another sort that is clear and thick, and there is a type as black as lacquer used in Chinese medicine. The remainder that is turbid and non-transparent is used only for glueing utensils. When painters use donkeyhide glue they dissolve it in pure water over a low fire and then only use the clear liquid on top.

3. Bottled liquid glue

This is ordinary glass-bottled liquid glue. In adding it to pigments, first it does not interfere with the pigments' inherent color; second, it does not itself produce any glare; third, it contains a preservative, so that during summer days it will not smell or curdle; and fourth, its strength in congealing and adhering is no weaker than that of oxhide or donkeyhide glue. I have used it for ten years now and have yet to experience any problems with pigments flaking off. In addition, it is very simple, convenient, and clean to use, only it is somewhat more expensive. Nowadays, there is bottled liquid glue with glycerine added to it, but it is definitely unsuitable for use.

4. Alum

Alum has an astringent taste and is produced by decocting and refining alum ore. It is translucent, much the same as quartz. The principal production site is Lujiang in Anhui Province. In nearly all types of Chinese painting, with the exception of ink-wash painting and large, freely done colored paintings, an alum solution is employed to fix the colors. If a painting has several layers of color applied, a light alum solution should be spread on, separating each or every other layer, especially the bottom layer of color. This is done in order to prevent the pigments underneath from spreading when the other colors are applied over them. For example, if a layer of cinnabar is put down first—no matter what concentration of glue it contains—and if no alum solution is applied, then when a heavy or light layer of rouge is washed over it, the cinnabar can begin to spread and to become mixed with the rouge. If alum is applied, the cinnabar will remain fixed no matter what kind of wash is applied over it.

In addition, to turn raw paper to sized paper, a solution of glue and alum must be brushed onto the surface of the raw paper (some names of raw paper are: *liu ji liao ban* 六吉料半 and *liu ji mian lian* 六吉棉連). Because ink and colors will halo and bleed when used on raw paper, glue and alum sizing are brushed onto it so that there is no haloing or bleeding and it is then called sized paper. Raw silk becomes sized silk—what we also call painting silk—by using the same method.

When we use painting silk to paint on, sometimes the silk to be used is damaged by dirt, or if a certain section of the painting does not seem quite right, or if what is written on it is found to be incorrect, then glue can be used to remove these problems and to restore the original white silk. The method for this is: using oxhide or donkeyhide glue, decoct it into a thick liquid glue and pour it onto the section that is ready to be restored. After waiting for it to dry naturally, grip the silk on the bias and stretch it tight. The entire layer of glue will crack apart bit by bit, and the section that you desire to remove will come up along with the glue, revealing the pure white silk ground. This method is entirely effective, as long as no grease, dirt, ink, or color has soaked through to the back side. The liquid glue that is poured out must be allowed to dry naturally; it cannot be heat dried or dried in the sun. It cannot be applied all over the silk, or the silk will develop cracks and the pigments will crack and flake off more easily.

Author's Notes for Chapter 1

A. "The Six Principles." In the Preface to his work, *The Old Record of the Classification of Painters,* the Southern Qi painter Xie He [ca. A.D. 475] wrote:

Although there are Six Principles in painting, rarely has anyone been able to work in complete accord with them, yet from antiquity down to the present day, individuals have been skilled at certain ones of them. These Six Principles—what are they? The first is: engender a sense of animation through spirit-consonance 氣韻 (one edition gives 運) 生動 . The second is: use the brush with the "bone-method" 骨法用筆 . The third is: fidelity to objects in portraying forms 應物象 (one edition gives 寫) 形 . The fourth is: conformity to kind in applying colors 隨類賦 (one edition gives 數, another edition gives 傅) 彩 . The fifth is: dividing and planning, positioning and arranging 經營位置 . The sixth is: transmitting and conveying [earlier models through] copying and transcribing 傳移模 (one edition gives 摸) 寫 (one edition gives 傳移摹寫). Only Lu Tanwei and Wei Xie have completely fulfilled these.[35]

"The Six Principles" comprise a famous ancient Chinese theory of painting. However, the various printed editions that have been handed down are not the same (e.g., the various Ming Dynasty printed editions in collected works and quoted in the publications of individual authors). The phrases are not the same (see above), nor are the interpretations identical (for example, the interpretation published by Zhang Yanyuan of the Tang in his *Record of the Famous Painters of Successive Dynasties,* or the explanation by Guo Ruoxu of the Song, published in his *Paintings Seen and Heard About,* and various later interpretations). What view should we take of this in the present day? This is a worthwhile problem for study in the research of Chinese painting. In order to facilitate research, the author has collated and compared the appropriate passages from each edition. The passage quoted above is according to the *Shuo fu* edition, as recorded in *Bai chuan xue hai.* The variant wordings found in each of the following editions have been placed in parentheses alongside each phrase so that they may be used as a type of reference material for researching "The Six Principles": the Ming Dynasty edition of *Record of the Famous Painters of Successive Dynasties, Wang Shih hua yuan, Jin dai mi shu, Peiwen Zhai shu hua pu, Mustard Seed Garden Manual of Painting, Gu jin tushu jicheng, Bai chuan xue hai, Yan bei ou chao, Meishu congshu, Zhongguo huihua shi.*

B. Zhang Yanyuan of the Tang, *Record of the Famous Painters of Successive Dynasties:*

When a workman wishes to be skilled at his work, first he must sharpen his tools. Plain silks from Qi and prepared silks from Wu, icy white silks and misty grey silks, fine and glossy and densely woven, these are the marvels of the loom and shuttle. (This refers to the silks used for painting.) Cinnabar from the wells of Wuling (Changde in Hunan Province), [cinnabar] sand from Mocuo (Jianyang in Fujian Province), sky blue from Yuesui (Xichang in Xikang Province), layered blue from Wei (Shanxi Province), flat blue from Wuchang (Zhang's original note: a high-quality mineral green), lead flower from Shu Commandery (Zhang's note: red lead), pewter extract from Shixing (Qujiang in Guangdong Province) (Zhang's note: barbarian white), [all of these are available] ground and refined, purified and selected, deep and pale, light and heavy, fine and coarse. There are yellows from Linyi (on the Malay Peninsula) and Kunlun, ant-ore from Nanhai (Guangdong Province) (Zhang's note: purple ore lac, from which pink and rouge are made), deer glue from Yunzhong (Shanxi Province), fish glue from Wuzhong (Jiangsu Province), ox glue from Donge (Shandong Province), and the juice of *Sagina maxima* (an herbal name, also called "Shu yang quan"蜀羊泉 , it is used in Chinese medicine)—refined and decocted, these were applied over pigments that were used in order to enrich them. (Zhang's note: the ancient painters all used the juice of *Sagina maxima.* Refined and decocted, it was called a 'color-enricher' and was applied in layers over green colors.) The ancient painters did not use 'first green' or 'pure blue' but

---

[35]This translation is adapted from Alexander C. Soper, "The First Two Laws of Hsieh Ho," *Far Eastern Quarterly*, 8 (1949), p. 423, and James Cahill, trans., "The Six Laws and How To Read Them," *Ars Orientalis*, 4 (1961), p. 380.

selected only the finest to accept for usage. Glue that is preserved for one hundred years (donkeyhide glue is used when it has aged, not when new) will not peel off in a thousand years (this refers to glue used together with pigments). With the hair of the giant eight-foot bamboo-eater [the panda?], each stroke is like a sword (this is about the brush).[36]

This work is from long ago in the year A.D. 847, and it is the earliest, most specific and most detailed record concerned with the use of silk, pigments, glue, and brushes. The sites for the production of pigments that it discusses include both local and foreign ones.

---

[36]This translation follows Yu Feian's wording and punctuation; cf. the translation in Acker, *Some T'ang and Pre-T'ang Texts*, pp. 187–90.

Chapter 2

The Development
of Chinese Painting Colors

From surviving cultural relics we can see quite clearly the development of Chinese painting colors, from an initial use of only pure, unmixed mineral and plant pigments, through a period of uninterrupted creativity, improvement, and step-by-step development. With progress came the use of mixed mineral pigments (such as chalk mixed with red to make a flesh tone, and azurite combined with chalk to get a sky-blue color) and mixed colors made from mineral and plant pigments blended together (such as indigo mixed with red to make a purple color, sophora yellow combined with malachite to get a light yellow-green color, and so forth). These kinds of combined mineral and plant pigments, in addition to the chemically manufactured lead white and red lead of antiquity, along with other foreign-imported pigments like rattan yellow and "purple stem" lac, were already extremely abundant by the time of the Southern Qi Dynasty in the fifth century and were used in pursuit of "conformity to the subject in applying colors." Passing through the sixth, seventh, and eighth centuries and the developments of the Sui and Tang dynasties, in the tenth century came the creation of the painting method where ink was used as a substitute for color.[1] From the twelfth century on, ink-wash painting and colored painting were on a par with one another, up until the early part of the fourteenth century, when the type of painting that employed mineral pigments such as azurite, malachite, and cinnabar was finally termed the "academic style" by the "scholar" painters, who said that it was not sufficiently refined. Painting done in thick colors in this period was already unable to claim equal status with ink-wash painting. From the time of this development on down to the nineteenth century, only the "chamberlains" [official painters] of the ruling class produced paintings done in thick colors. However, these colors are still in great use by folk artisans. Today, the Jiang Sixu Hall Pigment Shop [in Suzhou] specializes in the manufacture and sale of Chinese painting colors. Ordinary painters no longer need to manufacture colors for themselves.

---

[1]The development of monochrome ink-wash can be traced back earlier, to the 8th century at least, as the author indicates subsequently; for a history of this, see Shujiro Shimada, "Concerning the *I-p'in* Style of Painting," part 1, translated by James Cahill, *Oriental Art*, N.S. 7 (Summer 1962); Kiyohiko Munakata, "The Rise of Ink-wash Landscape Painting in the T'ang Dynasty" (Ph.D. diss., Princeton University, 1965).

The Course of Development of
Chinese Painting Colors

Let us first take a look at the pigments that were painted on ancient handicraft articles that have been handed down or discovered. We feel that they and the colors used in painting had a close relationship in their origin and development. Our ancestors, in the earliest age of man, had already discovered colors: for example, Zhoukoudian's Upper Cave Man employed a red color in dyeing ornamental articles, and the decorative patterns painted on neolithic painted pottery were done with chalk, red ochre, charcoal (colored black), and yellow ochre. Other examples are: brushmarks written in cinnabar and black on Yin [later Shang] Dynasty oracle bones, as well as cinnabar-painted grave goods. From this evidence, we can see how pigments were used in ancient times.[2]

The pottery and other objects from the Zhou and Qin dynasties (1143–207 B.C.) that we can see now, such as the painted pottery vessels of the Warring States period unearthed at Luoyang, employed red, yellow, blue, white, and black colors, painted in exquisite decorative patterns. On a pottery duck of the Warring States era excavated at Erligang in the city of Zhengzhou were bright colors of red, yellow, white, and black, with yellow used on the duck's beak and feet. Turning to literature, in the *Rituals of Zhou,* "Records of the Inspection of Works of the Winter Agency," it says that there are five types of officials in charge of painting on the colors. It also says that the painting of pictures is a matter of blending the five colors (the five colors are red, yellow, blue, white, and black). In the commentary to the "Great Chariot" chapter of the "Royal Customs" section of the *Book of Odes,* it explains that the coloring officials used the five colors to paint the banners and flags, jackets, skirts (the skirts being painted first and then embroidered), and so forth, of the ruling class of the Zhou. The principal use of color in the Zhou Dynasty and earlier was still on handicraft articles.[3]

---

[2]The oldest Chinese wall painting yet excavated, since the time Yu's book was written, is but a fragment that comes from the last reign periods of the Shang-Yin Dynasty (11th century B.C.; excavated in 1975 at Xiaotun, Anyang). The pigments found here included only black and red, not yet analyzed in terms of their substance, set on a lime-surfaced clay wall. Cf. *Kaogu,* 1976.4, p. 267.

[3]A trio of now well-known paintings from the Warring States period (ca. 4th century B.C.) have been excavated from two Changsha tombs, representing the three oldest Chinese paintings in color on silk: the so-called "Chu silk manuscript" and a funeral banner with a male figure riding a dragon were excavated in 1943 and 1973, respectively, from a tomb in the suburb of Zidanku, while a funeral banner with a female figure, dragon, and phoenix was excavated in 1949 in the nearby suburb of Chenjiadashan. The pigments used in these paintings have not yet been adequately analyzed and identified. The Chu silk manuscript includes black, red, purple, and grey-blue pigments. In the funeral banner from that tomb, while most of the work is painted in black ink, the figure of the man riding a dragon is said to have been done in color (these colors are not readily visible and are not identified in the report), while other parts of the painting include the earliest known use of gold and silver pigments. See Noel Barnard, *Scientific Examination of an Ancient Chinese Document as a Prelude to Decipherment, Translation, and Historical Assessment—The Ch'u Silk Manuscript,* Vol. 1 (Canberra: Australian National University, 1972), 18–22; *Changsha Chu mu bo hua* (Beijing: Wenwu chuban she, 1973); *Wenwu,* 1973.7, p. 4.

The first traces of Qin wall painting were excavated from palaces nos. 1 and 3 at the old capital, Xianyang, in 1974–75 and 1979, respectively. Archaeological reports from the former site indicate that among the approximately 440 small fragments, none quite as large as a square foot, were included black, red ochre, yellow, crimson, cinnabar, azurite, and malachite pigments, with black the predominant color followed by red ochre and yellow. These pigments were all derived from mineral ores, including lead and ilmenite (an oxide of iron and titanium). At palace no. 3, where larger

In the Han and Jin periods (206 B.C.–A.D. 417), literature and art flourished even more, and in the use of painting colors there was remarkable development. For example, the recently excavated Han Dynasty painted pottery vessels from Luoyang, in Henan Province, were painted with exquisite decorative patterns done in colors of red, yellow, azurite, malachite, white, and black. Also in the Han Dynasty tomb murals discovered in Wangdu County, Hebei Province, there are red, yellow, blue, green, black, and white colors. Moreover, painted there under the two figures of the "Chief Recorder" and "Chief Historian" are utensils that resemble an inkstone and ink. Comparing this work to the Han Dynasty tomb murals at Liaoyang and to the recently discovered tomb murals at Liangshan County in Shandong Province, it is somewhat more sophisticated in regard to color than they are. In the wake of increasing capital construction, there can be more discoveries of ancient cultural relics, which are the strongest material with which to document the development of colors.[4]

Paintings from the Western Jin Dynasty have yet to be discovered. The hand-scroll *Admonitions of the Court Instructress* painted by Gu Kaizhi of the Eastern Jin Dynasty (ca. A.D. 405, the oldest copy is in England, on exhibit in the British Museum) has for its principal colors red, red ochre, yellow, white, and black, while for supplementary colors it uses rouge, indigo, grass green, and sandalwood. Pottery vessels of the Jin Dynasty employing cinnabar, red ochre, yellow ochre, white chalk, and black charcoal are entirely in accord with this. This one scroll, regardless of whether or not it is a copy from before the Tang Dynasty, employs a

fragments were found, paintings were executed on a base prepared with clamshell white and included a range of mineral pigments, among them cinnabar, malachite, orpiment, and red ochre. Pale red ochre was used for the underdrawing, which was filled in with colored washes, some unmodulated, some irregular, and then redrawn with black lines. Although the paintings were originally done primarily in bright colors, a range of bright and pale colors was interspersed. Several of the colors appeared in two and even three different intensities, showing that the painters were already capable of separating their mineral pigments into varying color-layers through a liquid-separation and decantation (or elutriation) process. Cf. *Wenwu,* 1976.11, 23–24, and *Kaogu yu wenwu,* 1980.2, pp. 98–99.

[4]Since this was written, several Han excavations have provided the kind of materials Yu Feian foresaw, including the first Han silk paintings to be recovered. These include three funeral banners, far more skillfully colored than their Warring States period predecessors, two excavated at Mawang-dui, near Changsha, Hunan Province, and one at Jinqueshan, in Linyi, Shandong Province. An analysis of the now-famous banner from Mawangdui Tomb no. 1 (ca. 165 B.C.) by its copyist, Liu Bingsen, reveals that the pigments, bright and richly varied, included not only mineral but also vegetable and animal derivatives. Cinnabar (of more than one hue, some of it redder, some of it more yellow, indicating the use of an elutriation process), red ochre, and powdered silver (used both as a wash and in lines, but blackened with time) were its mineral pigments. Neither azurite nor malachite were used. The blues of the Mawangdui banner were derived instead from indigo (including flower blue), which in some places was tinted with white or mixed with other pigments to achieve a varied effect and a remarkable range of intermediate colors. Also vegetable in origin were its inks, used both for a pale underdrawing and a dark final redrawing after the colors had been applied. The white was derived from clamshells, lacking the original gloss of lead white but retaining its color well without blackening. Perhaps because it is such a coarse material, this clamshell white has peeled off in some areas. The Linyi banner included similar pigments: cinnabar, mineral yellow, clamshell white, indigo, and ink. Pale ink and cinnabar were used for a pale underdrawing, while some of the lines were later redrawn with cinnabar and white pigments (throughout much of the painting, the lack of outline is a surprising and most distinctive feature). A striking technique used in this painting for combining pigments comes in its occasional layering of semitransparent colors, mixing them, in effect, directly on the silk. For example, in one area a base of cinnabar is applied and then covered with a pale blue layer of indigo mixed with clamshell white, producing a purplish hue, most striking where the upper layer is thinnest in its application or has partly worn off. See *Wenwu,* 1973.9, pp. 74–75, and 1977.11, p. 31.

color method that has both principals and subordinate colors, and they are fresh and lively, vivid and strong.

It was in the period of the Northern and Southern dynasties and the Sui Dynasty (A.D. 420–617) that the "Six Principles" were imparted by Xie He. With regard to color, he put forward "Conformity to kind in applying colors," affirming the function and effect of color. However, paintings of the four dynasties of Song, Qi, Liang, and Chen (the Southern Dynasties), can only be known through the literary records and any authentic works still await discovery.

The greater part of the painting of the Northern Dynasties—the Northern, Eastern, and Western Wei Dynasties—that has survived to the present day can be seen at Dunhuang's Mogao Caves. Their color is characterized by skill in the use of green and blue. In the realm of color expression, there is a love of employing intense hues, powerful and bright, capturing the flavor of mountains and forests. The principal tones are those of mineral colors, while for the supplementary colors, rouge, indigo, grass green, and other plant pigments were used. Some of the intermediate colors included vermilion and red lead mixed with [lead] white. In the Wei Dynasty murals that are seen today, some of the figures like the female musicians and the flying apsaras have turned black because the vermilion and red lead mixed with [lead] white have changed over time. A certain number of the painters of that time, in their use of color, learned the method of achieving three-dimensional effects that was transmitted from India.

Colored painting during the Sui Dynasty gradually tended toward increased complexity and was greatly transformed. The Dunhuang murals illustrate this. Zhan Ziqian's painting, *Traveling in Springtime,* on exhibit at the Palace Museum's Hall of Painting [in Beijing], in its use of color opened up the way to combining ink and color, the method of applying color which is known as "color brushed on over ink."

In Tang Dynasty (A.D. 618–907) painting, the important formats were the mural, the hanging scroll, and the handscroll. Of the surviving murals that have already been discovered today, by far the largest portion is at Dunhuang, followed by Maijishan, as well as the recently excavated Tang Dynasty murals from Dizhangwan, at present day Xianyang in Shaanxi Province. [Mural paintings will be dealt with in some detail, shortly]. There are also scroll paintings still extant, to be discussed now.

The painters of the early Tang (A.D. 618–712) used both pure and mixed colors in their scroll paintings, which have survived more than a thousand years. The silk that we see is already spotted and disintegrating, yet the colors have by no means completely peeled off. Yan Liben's portrait handscroll, *Emperors of Successive Dynasties* (ca. A.D. 643) [Museum of Fine Arts, Boston], and Yuchi Yiseng's hanging scroll, *Portrait of the Heavenly King* [versions in the Freer Gallery of Art, Washington, D.C., and—attributed to Wu Daozi—in the Palace Museum, Taibei], illustrate that our ancient painters were not only skilled at utilizing color but also at putting their wisdom to use, assuring that the pigments firmly adhere to the silk so that through many years of rolling and unrolling they have not worn off, flaked off, or peeled away.

In the high Tang period (A.D. 713–765), owing to the great popularity of calligraphy in the Jin and Sui dynasties and after, the painters came under its influence. The inks of the time were dragon-incense-blend [a magical, Daoist ink made for Emperor Xuanzong of the Tang Dynasty], Zhen family ink, Yang family ink, and Wu family ink. Today we can see Wang Wei's *Snow on the River* [no longer extant] and the *Wang River Villa* [numerous versions exist], which all employ ink-wash, as well as the gold-and-green landscapes handed down since Li Sixun. Wu Daozi of the same era was known as the "sage of painting." Sometimes he used a light red ochre to fill in human faces and tree trunks, but other times he did not employ even red ochre, "rendering them only with ink lines." People called this "the Wu style." Still, his original works are no longer preserved, so we are in no position to assume anything about them. Among other paintings which have been handed down to the present time are Han Gan's *Pair of Horses* [Palace Museum, Taibei], Zhang Xuan's *Lady Guoguo on an Outing* (the copy by Zhao Ji [Emperor Huizong] of the Song Dynasty) [Liaoning Provincial Museum], Lu Lengjia's two portraits from *The Six Venerables* [Palace Museum, Beijing], Han Huang's *Scholars in a Garden* [Palace Museum, Beijing], and Zhou Fang's *Court Ladies Wearing Flowers* [Liaoning Provincial Museum], among others. Chinese painting, in the realm of painting method and in the developing use of color, had by this time already given rise to a considerable abundance of varieties and types.

The middle Tang (A.D. 766–820) to late Tang Dynasty (A.D. 821–907) saw not only the ascendance of landscape painting but also the rich and complex colored flower-and-bird painting. For example, Bi Hong painted pines and rocks on the walls of the Examination Hall of the Left (done in A.D. 767), Bian Luan was summoned to paint the new peacocks sent as tribute by the state of Luo (done between A.D. 785 and 805), and the Daoist Wuqiu Yuanzhi made a painting for Bo Juyi of lotus and lichee trees (painted in 819), among others. At this time, there were also landscape painters such as Sun Wei, Wang Xia, and Zhang Cao, and painters of flowers-and-birds, such as Liang Guang, Diao Guangyin, Zhou Huang, and others. Due to the uninterrupted development of painting itself, there came to be a division of labor among such categories as landscapes, flowers-and-birds, animals, and insects, all of those things that the people were fond of having depicted by the painters. During this period of time, paintings in ink and light color are estimated still to have been done in fairly small proportion. However, plant pigments were used by a great many. On the one hand, this is because mineral pigments in this period became brighter and deeper; but on the other hand, there was the gradual development of the lighter colors.

The ruling class of the Five Dynasties (Liang, Tang, Jin, Han, and Zhou, A.D. 907–959) and the two Song dynasties (Northern Song, A.D. 960–1127; Southern Song, A.D. 1127–1279) particularly valued the institution of a painting academy, selecting scholars to paint there. In this period, the painters came forth in large numbers, struggling in a battle to succeed. Chinese painting of this era underwent minute divisions of labor into ten classes: Daoist and Buddhist themes, figures, architecture, foreign peoples, dragons and fishes, landscapes, domestic and wild animals, flowers and birds, ink bamboo, and vegetables and fruits (see the *Xuanhe Catalogue of Painting*). While ink-wash painting constituted the largest proportion of work done, in the Five Dynasties, Northern Song, and the early Southern Song, the colored style of the late Tang Dynasty was still carried on, although it had already reached its creative peak. Some examples are: the *Night Entertainment of Han Xizai* [Palace Museum, Beijing] (for details, see Chapter 5), *Lotus and Golden Pheasant* by Zhao Ji of the Northern Song Dynasty [Palace Museum, Beijing] (with its lotus flowers of a light red color and its splendid golden pheasant clothed in the five colors), and *A Thousand Leagues of Rivers and Mountains* by Wang Ximeng of the Northern Song Dynasty [Palace Museum, Beijing] (using a dual-color application to paint the sky and water, employing both pure blue and pure green). The men of the "Hanlin Painting Academy,"**A** in their selection, preparation, and use of color can truly be praised for "letting a hundred flowers bloom." Wang Ximeng's *Thousand Leagues of Rivers and Mountains,* which was exhibited last year at the Palace Museum's Hall of Painting, even after so many years is still bright and clear, an excellent example of the use of color by an artist of the Painting Academy.

Paintings in color during the Five Dynasties and Northern Song Dynasty continued the style of the middle and late Tang. Paintings of Daoist and Buddhist figures gradually declined, while flower-and-bird painting and landscape paintings gradually developed. The same period also focused on brushwork and ink, emphasizing the "lofty resonance" and "spirit and bone-structure" of paintings (see the *Xuanhe Catalogue of Painting*). In Zhao Gan's *Early Snow Along the River* of the Five Dynasties period [Palace Museum, Taibei], Li Gonglin's *Five Horses* [no longer extant?], Wen Tong's *Bamboo in Snow* [Shanghai Museum?], and Zhao Ji's *Rare Birds Painted from Life* [probably referring to the work attributed to Huang Quan, in the Palace Museum, Beijing], of the Northern Song Dynasty, not only did each artist use his own type of line drawing, but at the same time each used ink-washes or color washes to give a more substantial appearance to the abstract images already produced. By the Southern Song Dynasty, ink-wash painting developed still further. Plain-line painting, which used ink-line drawing to bring out the forms of objects and ink and color for the washes, became more deeply respected. For example, the *Hundred Flowers* scroll on exhibit at the Place Museum and Zhao Zigu's [Zhao Mengjian's] *Narcissus* scroll [numerous examples] derive from pounce-copies [draft-sketches] of the late Tang Dynasty as developed through Li Gonglin's plain-line drawing manner.

In the Yuan Dynasty (A.D. 1279–1368), painting style was transformed; the ink-wash painting of the so-called "lofty men and recluse scholars" was widely practiced. Painting done in blue-and-green and heavily applied colors was denounced as "academic" by these ink-using literati and was no longer taken seriously. Therefore, during this ninety-year span the number of painters working in color could be counted on one's fingers, although there were several working in light colors among them. At this time, Li Kan (ca. A.D. 1318), in his *Bamboo Manual,* devoted several paragraphs to the method for employing pigments. He and Rao Ziran of the Southern Song Dynasty, in the statements in Rao's *Twelve Taboos of the Painting Masters,* both advocate that the use of color be delicate and subtly elegant. Further, Tang Hou, in his *Secrets of Painting,* said, "In mundane discussions of painting, it is invariably said that painting has thirteen categories." Tao Zongyi (of the late Yuan–early Ming period), in his *Zhuo geng lu,* records the thirteen branches of painting, the last branch being "carved blue and inlaid green." So we go from the few categories recorded in the "Economic Annals" of the *Sui History* and the "Arts Annals" of the *Tang History* to the "ten classes" of Song Dynasty painting and the "thirteen categories" of the Yuan Dynasty. Despite the promotion of ink-wash painting and subtle elegance by the so-called literati class, the people's favored "carved blue and inlaid green," which on the contrary is brightly colored, was entered as the thirteenth category. This category, moreover, was preserved down to the end of the Qing Dynasty. From this fact we can realize that the so-called mundane is really comprised of those things that we the people highly value, things that we have developed. There is extant today a volume used by specialists in "drawing portraits" (lifelike portraiture in figure painting), the *Technique of Painting in Colors* [by Wang Yi of the Yuan Dynasty].[B] This volume is an extremely important account from which we can examine the development of colors in Chinese painting.

We should have a look at the Dunhuang murals, which come entirely from the brush of folk artisans. The murals of the Northern, Eastern, and Western Wei—the Northern Dynasties of the Northern and Southern Dynasties period—can be called pure and fresh, robust and vigorous in their use of color, allowing those who view them to sense the spiritual beauty of the mountain wilderness. In the murals of the Sui and Tang dynasties, the colors tend toward the flourishing and sumptuous, the gorgeous and handsome, giving one a sense of abundant vitality. The Five Dynasties and two Song periods, following in the footsteps of the late Tang, stressed the blues and greens, the hues evoking a feeling of stagnation. Afterward, the standard became routine and there was very little stylistic creativity in color. In this period, only in the Yulin Caves murals of the Western Xia from Wanfoxia at Anxi do the colors still have a simple and unsophisticated style. The above-mentioned Mogao and Yulin Caves illustrate the special characteristics of the colors used by folk artisans of their respective periods.

As for the colors used in the Mogao Caves at Dunhuang, Mr. Xia Nai states in his research material that, "Altogether there are the following eleven types of raw materials: soot, kaolin, red ochre, azurite, malachite, cinnabar, lead white, red lead, indigo, gardenia yellow, and safflower (rouge). The first six are manufactured rather easily, as it is only necessary to crush them into a powder for convenient usage. The last five must undergo a rather complicated manufacturing process. This demonstrates that the Chinese people of that time were already able to make use of sophisticated techniques of pigment manufacture. In addition, these eleven raw materials for the most part are not local products of Dunhuang. Even in Dunhuang today it is not easy to get them all." (See *Wenwu cankao ziliao*, 2:5.) I agree completely with Mr. Xia Nai's views. For in actual usage, to merely grind material into a powder still was not sufficient. Now I will attempt to take the explanation of these eleven types of pigment a step further.

1. Cinnabar. This includes red standard and the vermilion made through ancient chemical manufacture.

2. Red ochre. This is divisible into three colors: palm brown, red ochre, and iron oxide. There is also "earth red," used in large quantities on Wei Dynasty murals.

3. Safflower (rouge). Rouge was made of safflower, madder, or purple ore lac; it was not always the safflower type.

4. Red lead. This is the same as lead yellow, which is also called Zhang red. There are two shades of it used on murals, one deep and one pale.

5. Gardenia yellow. In the Dunhuang murals that use a yellow color by itself, it is not certain if this is gardenia yellow or orpiment. Gardenia yellow is usually mixed with indigo or malachite. For example, in the Dunhuang Cultural Relics Research Institute's Cave No. 107 (Pelliot No. 050), the clothing of the late Tang female attendant figures is painted with mineral yellow. Orpiment is not to be used mixed together with lead white, so the "white ground" used for their clothing was not lead white but kaolin.

6. Azurite. The blue colors used on the Dunhuang murals are of seven different shades. Those of the Northern Wei had a deep blue type, but it is not known what kind of azurite it can be called.

7. Indigo. This is indigo blue, the selected essence of which is called flower blue or light indigo.

8. Malachite. There are five different shades of green used on the Dunhuang murals. One of them is verdigris, manufactured through chemical methods in antiquity.

9. Lead white. Mostly what was used there is "kaolin" and white chalk. That which was mixed with vermilion and red lead and has turned black is lead white, and there was also white chalk.

10. Kaolin. Also called porcelain clay, the best was produced in Qimen in Anhui Province. Its principal component is a hydrous silicate of aluminum (molecular formula: $Al_2O_3 \cdot SiO_2 \cdot 2H_2O$).

11. Soot. There were pine soot, lamp soot, and stove soot, which differ according to their original material.

The eleven types of raw materials listed above are not sufficiently comprehensive with regard to the research done on all of the colors used in the Dunhuang murals. Specifically, the most important type that has been left out is a red earth color used by large numbers of painters since ancient times—red ochre.

When the Ming Dynasty (A.D. 1368–1644) was founded, the Hanlin Painting Academy was set up on the basis of the Song Dynasty system. From then on, subsequent to the development of literati painting, we can only see that the early Ming painters Bian Wenjin and Dai Jin and Lü Ji, Qiu Ying, and others from the mid-Ming period were all skilled experts in the use of color but produced no new developments.

In 1590 (the eighteenth year of the Wanli reign), Li Shizhen recounted various types of pigments in his *Compendium of Medicinal Plants*. In 1637 (the tenth year of the Chongzhen reign), in his *Tian gong kai wu,* Song Yingxing also recounted various types of pigments, although neither of these was a book focused on recording Chinese painting colors. As for the colors used in Ming Dynasty print making, they were all colors that were used in traditional Chinese painting.

From the Ming through the Qing period (A.D. 1644–1911), there was some development in painting colors. Embroidered silks, lacquerware, porcelain, sleeves of clothing, uppers of shoes, New Year pictures, lantern paintings, colored paintings, prints, and others, all created a style of color unique to this period. At the same time, books containing several accounts related to color were printed and published, such as *Xiaoshan's Painting Manual,* by Zou Yigui, printed and published in 1756, and Ze Lang's *Painting Trivia,* printed and published in 1797, which detailed the methods of selection, preparation, handling, use, and so forth of Chinese painting colors. Despite the promotion of ink-wash and light color by the literati class of this time, the tradition of color in Chinese painting was highly valued, and up until the Opium War it was still vigorous in its development.[5]

[5] In 1956, Lu Hongnian wrote an article, "Notes on the Methods of Painting Walls in China" (*Wenwu,* 1956.8, pp. 15–17), which remains a valuable source of information today, providing among other things a chronology of how Chinese wall painters made successive improvements that permitted the pigments to adhere better to their base and to appear more lively, and which is worth summarizing here. Moisture was always a problem in painted walls, and for this reason, clay-core walls became far more prevalent than brick-core walls for use in Chinese painting, especially after the

Based on the situation recounted above, the selection, preparation, and use of Chinese painting colors has been in the hands of the individual painters all along (including among them the folk artisans), and how they did so was never publicly printed and published, especially so in the case of Imperial Painting Academy artists. Only after the early Qing Dynasty, as a result of social developments, did this information come to be printed and published, in greater or lesser detail. At this time, foreign colors also began to be assimilated and used. For China's tradition of painting in color, this was naturally a new era, an era of greater enrichment and growth.

Processing and Refining Colors
Assimilated from Foreign Sources

Pigments from the Western Regions and other foreign places, such as Malay Peninsula rattan yellow, Central Asian Hui or Moslem blue and granulated green, and the pure blue from Tibet and India, among others, were long ago imported into China through the Northwest and later imported by sea in large quantities.

Han Dynasty. The difficulties of the latter, Lu wrote, can be judged from the colors of the brick-core Han wall paintings at Wangdu. In the period from Han to Tang, most walls used for painting were composed of a single layer of rough clay made from yellow earth, or loess, mixed with wheat straw. Such a surface, even though rubbed smooth by hand, remained relatively coarse (affecting the brushwork, naturally) and was prone to rapid deterioration. At Dunhuang, beginning with the Tang Dynasty, a second layer of fine clay mixed with hemp fibers was added, and a coat of lime mixed with glue was brushed both between the two layers and on the outer, painting layer. While this provided a pure white layer for the artist to paint upon, the lime caused the glue fixative to fail and the attached material to break off. From the late Tang and after, this fixative came to be replaced by a mixture of sand and clay, the clay being pliable, which effectively secured both the second layer and the pigment layer to their base. This material failed only when the walls became wet and the sand lost its binding capacity. Moreover, because lime as used to prepare the surface of a painting (which can now be traced back to the late Shang Dynasty, see note 2, p. 22) was rough and hard to paint on, and because it became granulated and powdered off and blackened over time, it came to be replaced as a surface material by the use of white clay, the final layer of which was normally applied as a slip. White clay yielded a surface as white and smooth as plain paper, allowing much finer brushwork than lime, and it did not tend to "absorb the colors" as lime did (possibly, Lu suggests, because lime does not protect against moisture). The use of white clay for this purpose can be seen as early as the Wei Dynasty at Maijishan and Dunhuang and it eventually came into standard usage. The clay surface itself had to be covered before painting by three to four layers of a binder made from alum mixed with glue. Lu gives proportions—50 grams of glue to 100 grams of alum to 1¼ kilograms of water, varying with the season—but he does not provide any dates or sources for the introduction of this technique.

In the Five Dynasties, the painting surface was finished with a layer one- to two-tenths of an inch thick of lime mixed with hemp fibers, which was then rubbed to a fine finish and which produced enduring results. The Song architectural treatise, *Yingzao fashi,* Lu noted, contains a recipe for a surface-finishing material: 4½ kilograms each of sand, clay, and white-clay, twenty-five kilograms of macerated hemp (washed), and ½ kilogram of rough hemp. In the Yuan Dynasty, a new "wet-wall painting" technique was introduced, using sand and clay mixed with glue for the wall, which was smoothed while still wet and painted before it was dry so that the pigments could seep into the material and become fairly safe from decomposition as long as the moisture level of the wall was controlled. In the Ming and Qing periods, in order to secure the two layers of the wall still more effectively, large quantities of sheep and goat hair, cotton or hemp fiber, and paper pulp were added.

In addition, Lu's article offers interesting information on the materials and techniques traditionally used in transferring preliminary designs to the wall in Chinese wall painting.

There were also numerous painters who came to China and employed foreign pigments, foreign painters such as Yuchi Yiseng of the Tang Dynasty, Chadili [Kshatriya] of the Song Dynasty, Li Madou [Matteo Ricci] of the Ming Dynasty, and others, who all made a contribution to Chinese painting. After 1707, the Italian painter Lang Shining [Giuseppe Castiglione] came to China where he used Chinese pigments and studied the methods of Chinese painting. "Western red" or carmine was used from 1582 on (Zeng Qing [1568–1650] used Western red in portraiture). In traditional flower-and-plant painting, because Western red was so highly refined it has been very widely used down to the present time.

After the Opium War, foreign chemical pigments were imported in increasingly large amounts. By the first year of the Xianfeng reign (from 1851 on), foreign blue (made in Germany), foreign green (Grumbacher brand, made in Germany), and carmine (this carmine was made in Japan, England, and Germany in many types) were used everywhere in dyeing textiles, in the colored paintings done on architecture, and in folk artisans' painting. The reason for this is that they were inexpensive, gave good results, and were convenient to use. Thus, first the indigo-growing and indigo dye-manufacturing industries broke down, followed by safflower and madder cultivation. "Foreign blue" and "Grumbacher green" replaced azurite and malachite in painting architecture in color. By about 1920, folk paintings such as New Year's pictures were exclusively using carmine (also called magenta) and foreign green (also called aniline green). Besides the pigment that the Chinese painters could manufacture for themselves, the indigo that they could buy at the market was no longer indigo precipitate but a universally manufactured blue. Cakes of rouge were at this time already very difficult to obtain. The cosmetics used by women for adornment were all foreign products. At that time, the economic phenomenon of a semi-feudal, semi-colonial society of that era was fully manifest.

Chinese Painting Colors
Currently Manufactured and Sold

As a result of the various painting catalogues and painting manuals printed in the late Ming and early Qing periods which discussed Chinese painting colors, a shop especially for the manufacture of Chinese painting colors was set up in Suzhou, in Jiangsu Province, in the first year of the Qianlong era [1736]. This was the Jiang Sixu Hall, which has remained open until the present time. It was first established in "Wu Offices Eastern Scholars' State House" in Suzhou. Later, it expanded to Duting Bridge inside Suzhou's Chang Gate. Now it is at No. 32 Central Market, east of the Chang Gate.[6] In the beginning, the pigments that they manufactured and sold were excellent in their selection and finely prepared. They also mixed a glue solution with the pigments, producing pastes. When it was time to use them, a little water was added to make them dissolve, so they could be used very

---

[6]Once again relocated and now known as the "Suzhou Jiang Sixu Factory for Traditional-Style Painting Pigments," this famous institution is now at Tiger Hill Road, Bridge Number One, Boxian Branch Road.

conveniently. Aside from these, azurite, malachite, cinnabar, and other pigments were made into a fine powder which could be used by simply adding a bit of glue solution. Carmine still comes in small glass bottles, each bottle holding nearly one-half gram of carmine. Colors with glue pre-added are called "pastes," such as indigo paste, red ochre paste, carmine paste, and rouge paste. Pastes are further divide into "light glue," "Heaven" [top grade], and other grades. Because they have been meticulously manufactured, they can be sold nationwide. However, the indigo paste of recent years seems to have changed in quality; it is not good to use and exudes a foul smell.

In Beijing from 1911 on, some manufacturers have been able to produce azurite and malachite. The material selected for this malachite is somewhat finer than that from the Jiang Sixu Hall, and the standards of preparation are nearly the same. There is a clamshell white manufactured and sold in Beijing that does not change color and is whiter than lead white or zinc white. Since the Ming Dynasty, very few have been able to make and use this kind of white pigment.

Selecting the raw materials and manufacturing Chinese painting colors is extremely troublesome, so it is a real convenience for painters to have someone who makes and sells them; nevertheless, if Chinese painters do not understand the methods for the selection, preparation, and refining of Chinese painting colors and only rely on the manufactured and purchased types to use, then in doing detailed color paintings it will be difficult for them to make them bright, lively, and stable over time.

To sum up, from the development of Chinese painting we can see the development of Chinese painting colors. Before the Jin and Wei dynasties, unmixed mineral pigments were used as the chief colors and unmixed vegetable pigments used as supplements. Through ceaseless invention and improvement, from the Sui and Tang dynasties on, vegetable pigments, chemically produced pigments, and mineral pigments were combined for use. For example, rouge laid on over cinnabar becomes somewhat redder, cinnabar washed over indigo becomes more purple, and a light application of rattan yellow over malachite turns it to a delicate shade of green, while a light wash of rouge over lead white will yield a pale pink color. These colors washed over drawn and painted objects are plant and mineral pigments used in combination to produce "mixed colors" and "multiple-mixed colors." Following continuous invention, as mineral pigments were blended with other mineral colors, plant pigments mixed with plant pigments, and chemically blended colors mixed with other chemically blended pigments (such as vermilion or Zhang red combined with lead white to become a flesh color), not only were unmixed and mixed colors used together but in addition they were used for backpainting [painting certain supplementary colors on the back of the silk to bring out the brightness of the principal color in the front].[7] As this continued, by the Song Dynasty, in the painting of peony flowers, in addition to the painting of the backside the painting was also put through "three sizings and eight washes." (See the passage in Tang Liuru's [Tang Yin's] *Painting Manual*.) In other words, first a bottom layer of color was washed on, over that a layer of light alum solution was put down, then a third layer was washed on, and again over that a light alum solution was applied. After adding eight layers of wash, the color was

[7]For further comments on this, see Chapter 5, pp. 64ff.

adequate ("like a fragrant odor"); then over that one more layer of alum solution was put down, ensuring that the painting would never change color. From the Yuan Dynasty on, literati ink painting gradually gained ascendance, and as for investigations by painters into the use of colors, only the Academy members still grasped the methods of manufacture and application. In the middle of the Qing Dynasty, a special shop opened for the manufacture and sale of Chinese pigments, the Jiang Sixu Hall, which must be said to have facilitated things for the ordinary traditional-style Chinese painter.

Author's Notes to Chapter 2

A. The Hanlin Painting Academy. During the time of the Tang Dynasty, the ruling class had already established the official titles of "Painter-in-Attendance," "Painter-in-Waiting," and "Attendant" to try to ensnare the painters. In the Five Dynasties period, the Western Shu and the Southern Tang dynasties had both already established a "Painting Academy." The Song Dynasty further assembled painters, establishing the "Hanlin Painting Academy," and according to the relative superiority of the painters' artistic talent they were given the official rank of "Painter-in-Attendance," "Painter-in-Waiting," "Scholar of Art," "Full Scholar of Painting," "Student," "Attendant," and other official titles. In the Southern Song "Shaoxing Painting Academy," all alike were "Preferred" painters. However, there were some painters of the Northern Song who were unwilling to enter the "Painting Academy." Of the painters of Southern Song, nearly all were members of the "Painting Academy."

B. *The Technique of Painting in Colors* from *Portrait Painting and the Technique of Painting in Colors* by Wang Yi, literary name Sishan, of the Yuan Dynasty:

In coloring faces, you should first tone the ground using a mixture of third red [light vermilion], glossy white, and *fang* white [a white pigment, possibly in square tablet form], rattan yellow, sandalwood, yellow ochre, and Beijing ink. On top of this, make a thin wash using the white already used for the ground and then apply a wash of sandalwood and ink using a circular motion. If the complexion is to be white, then add a little yellow ochre and some rouge to the white, and if you do not use rouge, then add third red; if it is to be reddish, add a bit of red ochre to the previous color; if swarthy, add a little rouge to a mixture of white, sandalwood, and deep blue; if yellowish, add a bit of red ochre to white and yellow ochre; if black, add sandalwood, yellow ochre, and deep blue, a bit of each, to white. After the thin ground of white and the circular wash of sandalwood and ink, these are added on top, using more or less of them according to the clarity of the color, for these colors cannot all be treated identically. Around the corners of the mouth add some pale rouge, and if you want it to have a smiling expression let the two brushstrokes at the corners of the mouth rise up a bit. For the whites of the eyes, place a wash around the pupils in two strokes, then dot in the pupils using soot and draw the outlines in ink. If the eyes are slightly elevated [at the corners] and there are wrinkles, then they will smile. Go over the lips with a wash of rouge. If the nose is red, overlay it with a light wash of rouge. If the face is freckled, use a circular wash of pale ink; if pock-marked, use a circular wash of sandalwood. If dark-bearded, use a wash like that used for the hair; if purplish [bearded], use a wash-fill of sandalwood and ink; if yellowish-red, use a wash of yellow and sandalwood. For hair, first use a wash of ink, then use a wash of soot. You may choose from among the wash-fill, streaked wash, and disorderly wash. (These and the various washes mentioned above, circular, overlaid, *xuan*, covering, and *ran*, are all techniques used in painting with colors.) For fingernails, first use a wash of rouge, then apply a wash of white at the base. Whenever painting women's complexions, use white for painting the backside [i.e., painting on the reverse side of the silk, to make the colors on the front side stand out], then a thin overlaid wash of white, and then a wash of sandalwood and ink. The general rule for applying washes is this: on white paper, first is the *ran* wash, then a covering wash of white pigment, and after these another *ran* wash to bring things out; on silk, first paint the backside. (The above are comments with regards to methods of using colors for faces, fingers, and so forth.)

Here are all of the combined colors used in clothing and ornamentation:

-for deep red, use vermilion mixed with light purple;

-peach red, use vermilion mixed with rouge;

-jade red (in one edition, this is given as flesh color), use white as the base and add rouge;

-cedar green, add lacquer green [blackish green] to branch green;

-ink green, add shell blue [blackish blue, close to indigo with a reddish cast] to lacquer green;

-willow green, add sophora yellow to branch green;

-Mandarin green, this is the same as branch green;

-duck green [pale yellowish green], add saturated lacquer green to branch green; moonglow [a light blue], add blue standard [very light mineral blue] to white;

-willow yellow, add third green [light mineral green] mixed with a bit of rattan yellow to white;

-gosling yellow [pale yellow], add sophora yellow to white;

-for brick brown [grayish brown], add soot to white;

-for thorn-brown, add a combination of sophora yellow, shell blue, and the standard of yellow ochre to white;

-for artemesia brown, add a mixture of sophora yellow, shell blue, yellow ochre, and sandal-wood to white;

-for eagle's-back brown, add a mixture of sandalwood, soot ink, and yellow ochre to white;

-for silver-brown, add rattan yellow to white;

-for pearl brown [light yellowish brown] add rattan yellow combined with rouge to white;

-for lotus-fiber brown [lavender], add a combination of shell blue and rouge to white;

-for dew brown, add a small amount of yellow ochre mixed with sandalwood to white;

-for tea brown, add a combination of lacquer green, soot ink, and sophora yellow to a base of yellow ochre;

-musk brown, add soot ink to yellow ochre and sandalwood;

-sandalwood brown, add light purple to yellow ochre;

-mountain-valley brown, add the standard of yellow ochre [very light yellow] to white;

-withered bamboo brown, add a bit of sandalwood to white and yellow ochre;

-lake-water brown, add third green to white;

-onion-white brown, add the standard of third green [very pale mineral green] to white;

-birchleaf-pear brown, add a combination of yellow ochre and vermilion to white;

-autumn tea brown, add a combination of third green and sophora yellow to yellow ochre;

-ink-in-oil, add soot ink to light purple and yellow ochre;

-for the color of jade, add saturated third green to white;

-for camel color, add a bit of yellow ochre to white, the standard of lacquer green [light blackish-green], and ink;

-for fur pieces, add a bit of ink to white, yellow ochre, and sandalwood;

-for indigo, add saturated third green to third blue [light mineral blue];

-for golden yellow, add rouge to sophora yellow and white;

-for crow black, use a backpainting [painting on the reverse side of the silk] of Su blue [blue with a reddish or violet cast] and a covering wash of shell blue;

-for mouse-hair brown, add ink to yellow ochre and white;

-for not-deep red, combine light purple and vermilion;

-for grape brown, combine third green and light purple;

-for clove brown, add a bit of sophora yellow to a base of jade red;

-for apricot fuzz, add sandalwood to white, ink, and shell blue;

-for woolly damask, use light purple as a ground color and purple-pink for the patterns;

-for fur from the barbarian regions, combine yellow ochre with vermilion;

–for hides, use white as the base and light purple for the patterning;

–for otter fur, combine white and yellow ochre;

–for ivory tablets, use a good [amount of?] white and add a bit of yellow ochre and the white pigment will congeal;

–for black boots, use dilute soot ink;

–for *zhe*-wood [*Cudrania tricuspidata*] armchairs, combine white, sandalwood, yellow ochre, and soot ink;

–for gold-patterned *zhe*-wood, do the same as above, only do not add ink;

–for purple robes, combine third blue and rouge.

I am unable to record the others one by one, but you may select colors in direct relation to the objects depicted.

All of the colors used in combination can be carefully distinguished as follows: first blue [dark mineral blue], second blue [pure mineral blue], blue, third blue [light mineral blue], deep middle-blue, shallow middle-blue, shell blue, and Su blue, second green [pure mineral green], flower-leaf green, branch green, Southern green, oil green, lacquer green, red lead, pale vermilion, third red [light vermilion], red ochre, vermilion, branch red, light purple, rattan yellow, sophora yellow, peeled white [niter powder?], pomegranate seed, bright rouge, and sandalwood (for this sandalwood, add a bit of old ink mixed with rouge to vermilion).[8]

[8]Cf. the translation and discussion of Wang Yi's text in Herbert Franke, "Two Yüan Treatises on the Technique of Portrait Painting," *Oriental Art,* 3 (1950), pp. 30–32.

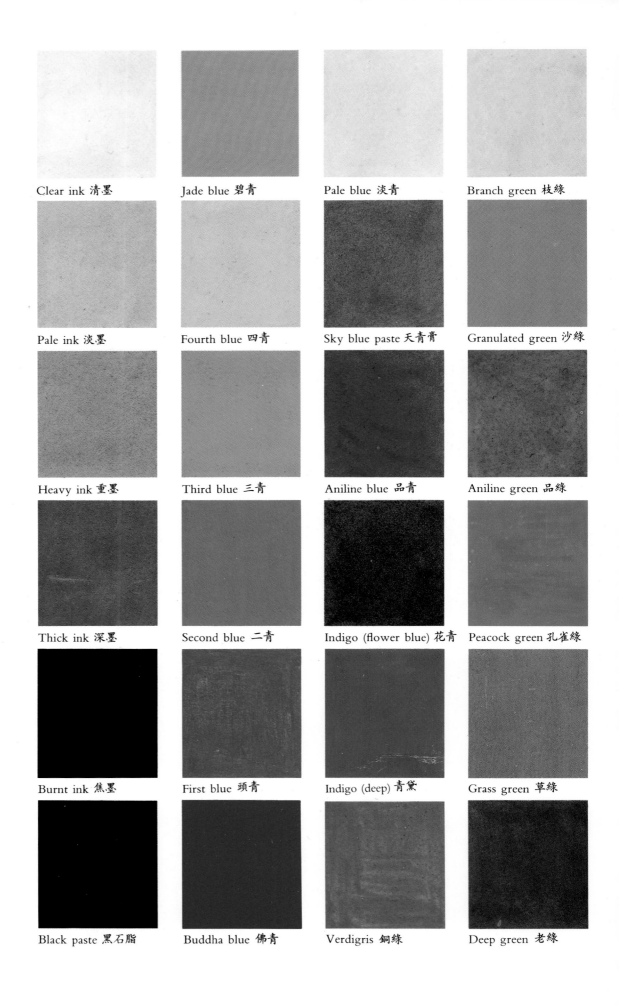

Clear ink 清墨	Jade blue 碧青	Pale blue 淡青	Branch green 枝綠
Pale ink 淡墨	Fourth blue 四青	Sky blue paste 天青膏	Granulated green 沙綠
Heavy ink 重墨	Third blue 三青	Aniline blue 品青	Aniline green 品綠
Thick ink 深墨	Second blue 二青	Indigo (flower blue) 花青	Peacock green 孔雀綠
Burnt ink 焦墨	First blue 頭青	Indigo (deep) 青黛	Grass green 草綠
Black paste 黑石脂	Buddha blue 佛青	Verdigris 銅綠	Deep green 老綠

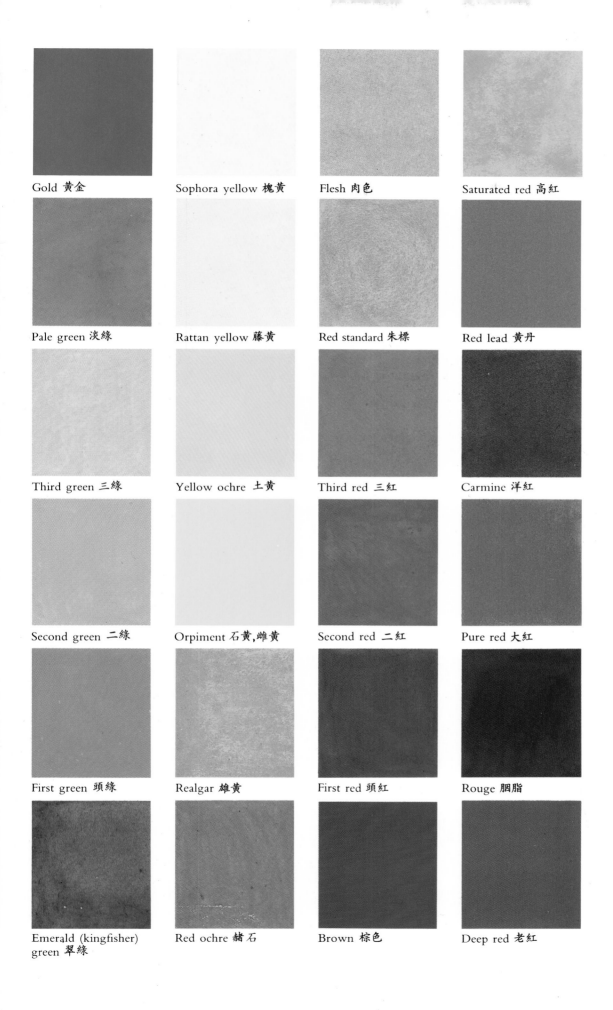

Gold 黃金

Sophora yellow 槐黃

Flesh 肉色

Saturated red 高紅

Pale green 淡綠

Rattan yellow 藤黃

Red standard 朱標

Red lead 黃丹

Third green 三綠

Yellow ochre 土黃

Third red 三紅

Carmine 洋紅

Second green 二綠

Orpiment 石黃,雌黃

Second red 二紅

Pure red 大紅

First green 頭綠

Realgar 雄黃

First red 頭紅

Rouge 胭脂

Emerald (kingfisher) green 翠綠

Red ochre 赭石

Brown 棕色

Deep red 老紅

Chapter 3
The Characteristics
of Chinese Ink

Chinese ink-wash painting is unique in character. Its special characteristic is the use of pale and deep, concentrated and light ink to express the tone and coloration of each of its subjects. Using ink that, on the whole, can be divided into five grades, "burnt" [or "scorched," or "roasted"], "thick," "heavy," "pale," and "clear," which were called the "colors" of ink in discussions of painting in ancient times (an explanation will be detailed below), ink-wash painting has had a unique position among the many kinds of Chinese painting.

The Chinese people's use of ink as a medium for calligraphy and painting originated very early. Whether or not ink was used on the late Zhou silk painting unearthed at Changsha we will leave aside for the moment, but other works such as the Han Dynasty slips inscribed in ink unearthed throughout the Dunhuang area, the "mummy" recently excavated near Turfan which has a reign title from the early years of the Han Dynasty written in ink on its white silk clothing, and the works in ink of the masters from the Jin and Tang dynasties on are the best evidence for the research of the problem of origins. The particular characteristics of Chinese ink are as follows:

1. By using a hair brush for calligraphy and painting, regardless of whether the hair is soft or stiff, flexibility of brush movements can be maintained without sticking or other hindrance.
2. Over a number of years' time, whether used on wood, silk, paper, painting silk, or other material, it does not deteriorate.
3. Over many years' time, the colors of ink do not fade or change.
4. If exposed to sunlight or heat, it will still maintain its black color.
5. When a painting is on paper or silk, although painted as fine as a gossamer, even if later stained by water, the ink will not spread or bleed, having a strong adhesive added for strength.

In the *Han Dynasty Rites of Officials* it says, "The Prefect of the Masters of Writing, his Supervisors, Assistants, and Gentlemen each month are granted one measure of Yumi large ink and one measure of small ink." Yumi is a place name, in modern day Qianyang County, in Shaanxi Province, a production site of pine soot and ink in the Han Dynasty. In the *Shuo wen,* written by Xu Shen in the Later Han Dynasty, it says, "Ink is black; it is a dust made from pine soot." In the late

35

Han and Three Kingdoms period Huang Xiang discussed ink, saying, "Much glue makes the ink black." From these ancient literary records, it can be proven that more than two hundred years B.C., in the Yumi region, pine soot was already used to manufacture "large pieces" and "small pieces" of ink. Xu Shen simply talked about dust made from pine soot. Only with Huang Xiang was it explained that ink was made into "large pieces" and "small pieces" by means of "much glue."

Jia Sixie of the Eastern Wei (A.D. 6th century), in his *Qi min yao shu,* set forth a method of ink manufacture. This is the earliest work that records how the working people made ink. In this work it says, "... Pound up good, pure soot, and by means of a strip of fine silk placed over a jar strain out the vegetable material, so that what remains is like fine sand or dust. This material is very light and fine, so you should not expose the strainer or it might fly out and be wasted. You must be cautious. One-half kilogram of ink (by which he means pure soot) takes 250 grams of good glue soaked in the sap of ash tree bark—this ash tree is the Jiangnan *fanji* tree [*Fraxinus bungeana*]. This bark is a light green color; it thins out the glue, and in addition it enriches the color of the ink. One can also add five drops of eggwhite, discarding the yolk. Taking 50 grams of real pearl [powdered] and 50 grams of musk, put them through a strainer separately, then mix them together. Mix them in an iron mortar; it is best to keep them firm and not to moisten them (in other words, it is better to keep them thick and not let them become dilute). Pound them 30,000 times with a pestle; the more it is ground the better. Do not mix the ink in the second month or the ninth month, for when it is warm it will spoil and smell, but if it is too cool it will be difficult to dry out. ..."

From this passage, one can make out that the process of creating ink was extremely complex and was the fruit of labor and creativity. From the Tang Dynasty on, there have been changes in the "secondary ingredients" for the making of ink (soot acts as the principal ingredient, with several kinds of herbal materials as secondary ingredients). In the Tang Dynasty, ash tree bark, Chinese honey locust [*Gleditschia chinensis*], chalcanthite [or blue vitriol], verbena, vinegar, pomegranate bark juice, powdered rhinoceros horn, rattan yellow, and croton bean were used. In the Song Dynasty, not only was pine soot used, but oil-soot lampblack was especially used in making ink (one-half kilogram of tung oil, when burned, will yield more than 50 grams of "top soot").[1] (From the Five Dynasties on, the manufacture of soot was transferred to Huizhou in Huangshan.) Besides the above-mentioned "secondary ingredients," raw lacquer, oxhorn marrow, pig gall, carp gall, white sandalwood, clove, Borneo camphor [also called "dragon-glue"], garden burnet, gallnut, the rhizome of Chinese goldthread [*Coptis chinensis*], Asian puccoon [or "purple-herb," *Lithospermum officinale*], madder root, black soy bean, logwood, walnut, root of Chinese monkshood [*Aconitum carmichaelii*], tree peony bark, linden seeds, indigo, cinnabar, and other "secondary ingredients" were also added. There is no record of the kinds of "secondary

---

[1]Pine soot was produced from pine wood burnt in kilns, either vertical kilns with a series of stacked, perforated jars in place of a chimney, in which the soot was deposited (the crudest soot filling the lower jar, the finer quality soots progressively filling the upper jars), or horizontal kilns with long exiting flues leading to small, closed cells where the soot gathered on paper-lined walls. Oil soot was produced by burning small, cup-sized earthenware lamps of tung or other oils, a small number of them placed in a larger water-filled bowl and each of them partially covered by rounded, inverted, carefully polished lids in which the soot collected; the lamps were lit with grass or vegetable-fiber wicks.

ingredients" used in the Yuan and Ming periods, only that in Cheng Junfang's manufacturing ink the kinds of "secondary ingredients" were reduced to fifteen or sixteen. In the Yuan and Ming periods, oil-soot lampblack was considered most important for the manufacture of soot. Aside from tung oil, vegetable oil (oil made from the Chinese tallow tree of Jiangnan [*Stillingia sebifera*]), sesame oil, and lard were used. Also used was spoiled lacquer burned for soot, called "lacquer soot." (The above comes from Jia Sixie's *Qi min yao shu* of the Eastern Wei period, Chao Yidao's *Ink Manual* from the Song Dynasty, Li Xiaomei's *Ink Manual* from the Song Dynasty, Shen Jisun's *Gathered Essentials of Ink Methods* from the Ming Dynasty, and Song Yingxing's *Tian gong kai wu* from the Ming Dynasty, as does the following.)

There is good and bad soot—regardless of whether it is pine soot, oil-soot lampblack, or lacquer soot—and all soot is made by burning in a kiln. The soot burned near to the fire is called "body soot" and is of inferior quality; that from the center of the kiln is called "neck soot" and is of a middle grade; and that at the farthest distance from the fire, on the four sides and at the top of the kiln, is called "top soot," "head soot," or "best soot," which is first-class pure soot, of the best quality. When we buy ink today, the tops of some ink sticks are inscribed "Pine Soot" and some "Lacquer Soot," representing the difference in the raw material of their soot. Some have written on them "Best Top" or "Best Soot," some have "Tribute Soot" ("tribute" means the best quality soot that was used to "pay tribute," given to the emperor for his use), and some have written on them "Superb Tribute Soot." These are all ways for the ink manufacturers to say that their ink is made from the best quality soot. The next best are marked "Select Soot," but there are certainly none marked "Body Soot" or "Neck Soot." The ink currently sold by the Hu Kaiwen–Cao Sugong Inkshop in Huizhou still employs "Best," "Top," "Tribute," and "Select" as marks to indicate the quality of the soot used in the ink.

One other important component of ink manufacture is donkeyhide glue, oxhide glue, and other animal glues. Specialists in ink-making through successive dynasties all advocated using aged (Zhang Yanyuan of the Tang Dynasty said that "100-year preserved—donkeyhide—glue" should be used to add to pigments), clear, and light animal glues in making ink. If newly made glue is used, and if eggwhite, ash tree bark, clove, root of Chinese monkshood, or pomegranate tree bark acid cannot be added, then when this kind of ink is made it will easily warp, crack, and smell bad.

How We Choose and Purchase Ink for Use

In the ink-making of the Qing Dynasty, the ink workers increased the number of times the ink was ground with the pestle to "100,000 strokes" and reduced the number of "secondary ingredients" (just which "secondary ingredients" were eliminated is unclear). At the same time, the government soot factory was established (under the "official title" of "Jiangning Imperial Silk Factory," which was revoked only in the first year of Guangxu's reign [1875]). As a result of my few brief decades of testing ink, I feel that the better "imperially manufactured inks" include only "Inner Palace Light Coal," "Black Jade Ring," and "Imperial Poem on the Plowing and Weaving Pictures Ink" (all names of ink) of Xuanye (the Kangxi Emperor), and the remixed ink (selected Ming Dynasty ink crushed and mixed with new soot and reprocessed) of Hungli (the Qianlong Emperor); however, for use in painting, one must still employ ink of the Tongzhi Emperor or the first years of Guangxu's reign. Both types must be ground to be used, and in using ink for producing traditional Chinese painting these days one must first take note of the following points:

1. Before Liberation, due to imperialist invasions, large amounts of lacquer and tung oil (the raw materials for making soot) were exported, while at the same time American "gas soot" was imported, which replaced the pure soot produced in China. It is composed of 90–95% charcoal and its color is superior to pure soot. After 1880, makers of ink gradually came to use it. This kind of ink is suitable for doing freely done painting on raw paper, but if used for precise outline-and-wash on sized paper or silk it will bleed or halo easily.

2. The Hu Kaiwen–Cao Sugong factory and others still currently manufacture and sell ink. The sale prices are fixed according to the weight and the ingredients in the soot. They also have new ink fashioned according to old-style molds.

3. When you select and buy ink, you should first look to see if it is fine, smooth, moist, and glossy, then to see if it is suffused with a glow of blue or purple, for ink that is not suffused with a blue or purple glow is second-rate. After that, look at the age and authenticity of the "gold-rinse" and "blue-fill,"[A] then consider the shapes and decorative patterns. If you see that it is not carefully made, that the wood grain of the mold is visible, that it is not velvety, moist, and glossy, or has a light and rough feeling when stroked, then it certainly has been manufactured in a rough and slipshod way and is a substandard product of inferior material and shoddy work. Next in importance is whether or not there is a period date and the name of the maker of the ink [molded onto the inkstick].

4. Painting is different from calligraphy in that the ink used in writing need only be black and moistened for use; painting, especially ink-wash painting, finely drawn figure painting, and flower-and-bird painting, demands above all that thick ink be truly thick, that black be a true black and not suffused with grey, and that pale be truly pale. If its fixative strength and adhesive force is great, when ink is washed over with colors or water it will not bleed, halo, or run. Therefore, old ink that is too many years old is not suitable for these special uses. With some old ink the glue has lost its efficacy, or sometimes it has gotten damp and become suffused with a greyish-white powder.

5. The recent ink made from foreign soot, in terms of the blackness of its color, is somewhat blacker than ink manufactured from other types of soot. However, with regard to its light color, it is not possible to make a clear tone which has expressive spirit—"penetrating to the bone." The reason is probably because the number of strokes with which the ink is ground has been reduced.

Of the inks prepared for painting which I have studied, one type is pine-soot ink (made by Cao Sugong, literary name Danyou), one type is lacquer-soot ink (made by Wang Jiean), and one type is oil-soot ink (made in the Qianlong period). Oil-soot ink made from lacquer soot mixed with oil soot was made by Hu Kaiwen in the Tongzhi period. Pine soot was selected for its dark, lusterless black and was used in painting feathers-and-fur and for painting butterflies; lacquer soot was selected for the luster of its dark black color and is used to dot in the eyes. These two kinds of ink, when mixed with oil-soot ink, become especially black and glossy, able to stand up to washes without bleeding or haloing; they can be thick or pale, and their clear grade is able to "penetrate to the bone" with expressive spirit ("penetrate to the bone" means to soak into the paper).

Essential Information about the Proper Use of Ink

What is recounted in this section is information relating to the use of inkstones, to water, the washing of inkstones, and "overnight ink," as well as to the ink in ink-wash paintings. I will divide the discussion into the following sections.

1. Inkstones

Inkstones and ink have a direct relationship. The two types of well-known ink-stone materials are Duan stone (which comes from Duanxi in Guangdong Province) and She stone (which comes from Longweixi in Shexian, Anhui Province), both kinds of which are aqueous rock and comparatively easy to grind the ink upon. Duan stone is not as good as She stone. In the time of Li Yu of the Southern Tang Dynasty (A.D. 961), She stone was already being mined and manufactured by the laboring people. We must first require of a stone that it draw the ink down [into the stone]. We also require of an inkstone that no matter how hard the inkstick, when we grind it up it should be easy to grind into a concentrated state while at the same time there should not be bits of stone floating up as a result of the ink being ground upon it, and this constitutes a good inkstone. I consider that the She inkstone most convenient to use is the type known as "ink sea." It is made in square and round forms, large and small sizes, with a stone cover on top and a stone spout on the side. It is carved entirely from a single block of She stone. If one is using a lot of ink, it is ground in the reservoir of the sea; to use less ink or a different kind of ink, the stone cover can be used for grinding the ink, which makes it extremely convenient to use. Mr. Qi Baishi is one painter who used a She stone "ink sea" to carry out his creative work. Both brush shops and ink shops carry this kind of ink sea.[2]

2. Water

The best water to use for grinding ink is bitter and astringent-tasting well water. It can "bring out the ink"—that is, cause an increase in the degree of blackness and the luster of the ink; moreover, its use in painting somewhat improves the adhesive strength of the ink. It is also somewhat better if the water used for mixing colors is well water or another kind of natural water. Water that has been disinfected with chemical reagents such as bleaching powder, however, is not suitable for use.

[2]For more on inkstones, see *Masterpieces of Chinese Inkstones in the National Palace Museum* (Taibei: National Palace Museum, 1974); R. H. van Gulik, *Mi Fu on Ink-Stones* (Beijing: Henri Vetch, 1938). For Duan and She stones, see Liu Yanliang, *Duanxi ming yan* (Guangzhou: Guangdong renmin meishu, 1979), and Mu Xiaotian, *Anhui wenfang sibao shi* (Shanghai: Shanghai renmin meishu, 1961).

3. Washing the inkstone

Inkstones must be carefully washed. When fine painting is to be done on sized paper or silk, before the ink is ground one must first thoroughly clean out any ink from the reservoir of the inkstone. If it is not washed clean, the colors of ink will not emerge, and when [this ink, on the painting] is washed over with water it will easily bleed and halo, run and come off. What is "overnight ink"? It is an ink solution that has stood for one night and returned to a dry state. Painters of ink-wash landscapes all like to use this "overnight ink" since it is easy to apply in washes. However, in the summertime and in the south, "overnight ink" that has stood all night is already no good for use and even gives off a foul odor.

4. The colors of ink

In ink-wash painting, the colors of ink are divided into five categories of tonal intensity, which are burnt, thick, heavy, pale, and clear. Here it is first necessary to consider the quality of the essential character of the ink used. Ink or refined-ink made from American "gas-soot" (produced in Shanghai, some of it marked "Heaven," some "Long Life") yields only burnt and thick tones, while pale tones are difficult and there is no need to discuss whether its clear tone has any expressive spirit. The inks of the Tongzhi and Guangxu periods yield only the burnt, heavy, and pale tones and neither a thick black nor a clear light one. Generally speaking, the five colors of ink are as follows:

1. Burnt ink. This is a ground ink solution which has evaporated for half a day in the inkstone reservoir and is then used in painting for the deepest and most prominent details. These are the darkest sections of the whole painting, black and glossy.
2. Thick ink. That is to say that its degree of blackness is next to that of burnt ink. Burnt ink can be glossy, but thick ink has a greater liquid content so it is not glossy even though it is black.
3. Heavy ink. This is distinguished from pale ink; its liquid content is somewhat greater than thick ink, and it looks somewhat darker than pale ink.
4. Pale ink. The liquid content is increased to create a grey color called pale ink.
5. Clear ink. With regard to the colors of ink, this is a shadow with just a bit of light grey color, a shadow that displays hazy forms like those of morning fog and evening mist.

Generally speaking, good ink is able to produce not only burnt and thick but pale and clear tones as well, determined by the number of times that the ink is ground when it is made. As for painters who use the pale tone of ink in doing ink-wash paintings, this is really no easy matter. Among the painters of antiquity who had sufficient skill in the use of ink to bring out the pale color of ink, Ma Yuan (*hao* Yaofu) of the Song Dynasty, Fang Congyi (*hao* Fanghu) of the Yuan Dynasty, Yun Shouping (*hao* Nantian, *zi* Zhengshu, born in 1633) of the Qing Dynasty and Hua Qiuyue [Hua Yan] (*hao* Xinluo) of the Qing Dynasty were painters who truly excelled at the use of ink. The landscape paintings and flower-and-bird paintings of these masters which can be seen today are still bright, and their pale tones have expressive spirit. Although this is not exclusively due to the intrinsic quality of the ink, it is still directly related to its inherent character. Moreover, among the colors of ink, a clear tone is spiritually expressive, because it accents the other dark and pale shades of ink color. In other words, in the empty sections of a painting, certain elements are painted as clear and pale shadows, and these clear, pale shadows have an extremely important expressive value within the painting, as they reflect the specific character of a certain moment in space and time, causing those who see them to penetrate into the painting and to visualize many elements lying hidden in these shadows. Therefore, we consider this kind of painting to be inexhaustible in its appeal to the imagination.

5. Grinding

Grinding (or ink-grinding) is an important task for each painter. When painters and folk artisans teach their apprentices, they teach them how to grind the pigments first, as a matter of principle. They rise early in the morning and always grind up a reservoir full of ink first, preparing it for when it will be used in creative work. So, it is better to say grinding the ink than grinding the man, of an activity that exercises wrist strength, the arm, and the elbow. [An old adage says, "It is not the man who grinds down the ink but the ink which grinds down the man," meaning to be worn out by literary activities.] When the days [of the year] lengthen, this becomes increasingly helpful for painting. In his early years, Mr. Qi Baishi always ground the ink himself after he got out of bed, according to this same principle. The method of grinding ink is to go especially slowly when you first start, not grinding with too much force, so that the dry, solid inkstick will not break or produce an inky sediment because it is ground too rapidly or ground too forcefully and crumbles on contact with the surface of the inkstone. If water is applied directly to the inkstick at the start, then with slow, light grinding the moisture will soak into the part of the inkstick that comes into contact with the inkstone and make it rather soft. Later on, increase the speed and force of grinding and it will be easy to grind up a thick ink solution, saving time and avoiding any sediment. Ink from before the Qianlong period (A.D. 1795) requires this method all the more—grinding slowly at first and gradually increasing the speed to avoid crumbling.

6. Storage

With ink, one mostly has to fear dampness and wind. When ink gets damp the glue is ruined, although the soot is not harmed. The wind will make it crumble easily, especially newly manufactured inksticks. To store ink, it is essential to wrap it up in paper and put it in a comparatively dry place. Good ink that has already gotten damp, if it is mixed together with new ink, ground up and used together with it, can still have occasional use in painting, although it will be black and without luster. But even better is to use the original box for storage. All of the ink used by the people is new ink. They take the new ink they have bought, wrap it tightly in paper first, then expose the lower part where they will grind the stick, about two-thirds to a whole centimeter long, and they melt wax evenly over the portion in the paper, and whenever necessary they use a knife to break off the wax and the paper [from the inkstick]. In this way, they can keep the inkstick protected forever from crumbling. If an inkstick is broken, in order to put the fragments back together again use the method given above—line the outside with paper, melt wax over the surface, and then it can be used as a solid stick.[3]

Supplementary Note: Colored Inks

["Colored inks" are very high quality colored pigments that are molded into the shape of inksticks.] Colored inks come in sets of five colors and ten colors. For example, in the "Five Fragrances" manufactured in the Qianlong period, the five colored inks are azurite, malachite, cinnabar, orpiment, and white; in the "Ten Famous Flower Friends" manufactured during the Jiaqing period, the ten colored inks are cinnabar, orpiment, azurite, malachite, "mother-of-pearl white," purple ore lac, red lead, realgar, red ochre, and red standard.[4] These were originally used for writing critical comments on literary texts. In using these colors for painting pictures, not only are they composed of good raw materials, but the glue that is used with them is also especially good. The method of handling them does not require that they be ground up on an inkstone, but rather crush them first, place them in a bowl to which water is added, and steam them in a steaming pan until there are no longer any lumps; then add some boiling water, mix it up evenly, and leave it to settle out. After approximately two or three days, the glue solution will have floated up to the top. Skim this [glue] off and immediately dry it in the sun or with heat, preparing it for when you will add it back to the [painting] colors for use. This is an extremely good, light glue. The remaining color should then be reground and precipitated again.

Author's Note for Chapter 3

A. Gold-rinse and blue-fill. Gold and azurite [inscriptions on cakes or sticks of] ink can reveal its age. The greater the age, the more the gold will be suffused with red and the more vivid the blue will become. This comes from a special study of gold and azurite made by an eminent producer of the ink.

---

[3]For further information on the history of Chinese ink, see Berthold Laufer, "The History of Ink in China," in Frank Wiborg, *Printing Ink: A History* (New York: Harper and Brothers, 1926), 1–52; Herbert Franke, "Kulturgeschictliches über die chinesische Tusche," *Abhandlungen, Bayerische Akademie der Wissenschaften,* 54 (1962), pp. 3–158.

[4]See *Masterpieces of Chinese Writing Materials in the National Palace Museum* (Taibei: 1976), pls. 9 and 13, for color illustrations of 5- and 10-color ink sets.

Chapter 4

Folk Artisans' Use of Colors

When folk artisans make portraits, painted sculpture, colored paintings, New Year's pictures, and lantern paintings, they use colors that are bright and lively, bold and exaggerated, and which can be completely enchanting whether seen from near or far away. They demand of color that it be "sharp" and "bright" (see the following sections), and they have succeeded in creating a unique style of color.

In the old society, folk artisans occupied an exploited social status. At the same time, the master artisans still exploited their apprentices. Because they carried out their work under contract for both materials and labor, they had to apply careful calculations and strict budgeting to ensure that their product was good and that it would be easy to distribute. For their colors, they carried out precise research and analysis on which kinds were easy to obtain, which kinds were inexpensive yet gave good results, which kinds were convenient to use, which would be relatively opaque, and whether they remained unchanged over time, and only after this was it decided which kinds of materials were to be used.

Looking again at the Dunhuang murals, comparing the colors used there (for details see Chapter 2) to the colors used by the "artists of the Central Plain" [i.e., from the northern metropolitan centers] from the Southern Dynasties, the Sui, Tang, Five Dynasties, and Song, it is clear that there are differences in their cost and quality, among other things. (For the Central Plain artists' use of color during the Tang Dynasty and earlier, see Zhang Yanyuan's account from his *Record of the Famous Painters of the Successive Dynasties*—in our Chapter 1 [author's note B]; there was further gradual development after the Tang). There are several Dunhuang murals in which the master workers drafted the faces, the apprentices applied the color, and the masters then drew the [final, visible] linear outlines. On murals that have already peeled away, we can still see this as clear proof of the lines (taken from the poem by Mei Shengyu [Mei Yaochen] of the Song Dynasty on Xu Xi's painting), "Deep in years, peeling white reveals the traces of ink." On the Tang Dynasty murals excavated most recently from the tomb at Dizhangwan in Xianyang, Shaanxi Province, the outlines of ink brush painting are completely visible, for the colors are filled in following alongside the ink lines. These murals employ the "outline-and-fill" technique of the "Central Plain painters." Also, with the "Central Plain painters," often the filling in of colors in outline-and-fill painting came entirely from the hand of one person, while with the folk painters this is not certain.

The Use of Color in Portraiture and Colored Painting

The colors used by the folk artisans of the past in the paintings and sculpture of Buddhist and Daoist images of deities and worshipers and in the colored paintings done on architecture were part of the same tradition of types and techniques as the art of Dunhuang. There are obvious differences between their use of color and that of the literati artists. In colored painting, artisans stressed "piled gold and dripped white" (dripped white is also called "standing white");[1] they stressed outline-and-fill and also the technique of dipping one brush into a number of colors in order to paint one object with several hues, which creates the feeling that one is witnessing the emergence of solid form.

According to comments made by a Beijing producer of colored painting, Liu Xingmin, they employed the following colors:

1. Zhengshang vermilion. (Zhengshang is a trademark. This is a vermilion used in lacquer in China, not easy to find.)

2. Sunlight vermilion. ("Sunlight" is a trademark.)

3. Saturated red. (This is the finest pure red, a foreign color.)

4. Guang saturated red ochre. (This is a superior red ochre, produced in Guangdong Province.)

5. Carmine. (This is available from Jiang Sixu in bottles of one-half gram. It is becoming increasingly expensive.)

6. Foreign red pearl. (Its color is bright, it is inexpensive, and it can be used as a substitute for carmine.)

7. Red ochre. (This is made by the painters themselves.)

8. Zhang red [red lead]. (This is made in China [in Zhangzhou, Fujian Province], in packets of 500 grams.)

9. Guangming red. (Guangming ["Bright"] is a trademark. There are also red lead, "pottery red," and "barrel red," which are somewhat inferior.). [Guangming is considered the finest quality.]

10. Orpiment. (That which is made in China is not as good for use as the French, the Chinese being both expensive and foul-smelling.)

11. French orpiment. (This can be separated into three colors, from deep to light. It is both good and inexpensive.)

12. Moon yellow. (This is rattan yellow. After it is used, the leftover lumps will turn hard and red and produce a sediment, but if it is steamed thoroughly in a pan and cold water is applied to "cool" it then it can be used.)

13. Realgar. (This is made by the painters themselves.)

14. Shunquanlong Buddha blue. (This is the Shunquanlong Shop's brand of Buddha blue. The character for Budda 佛 can also be simplified to "*fu*" to be written as 伏.)

15. Mao blue. (A foreign color, it is also called deep indigo. It is a deeper color than Buddha blue and can be used as a substitute for light indigo.)

---

[1]These refer to techniques for using gold and white pigments in relief. For pigments to be piled, or mounded, means that a broad surface area of pigment has been raised above the ground level, as with gold leaf, while dripped white, or standing white, refers to raised lines of pigment; today these terms have become merged. See Lu Hongnian, *Wenwu,* 1956.8, p. 16, and further discussion below, p. 46.

16. Grumbacher foreign green. (By grinding and precipitating it, both deep and light can be obtained. It is a product of Germany.)

17. Chanchen foreign green. (This is a foreign product from the Chanchen [?] Company in Germany. There is also a Qianxing foreign green.)

18. Emerald green. (A foreign color, also called "bright [*geba*] green." These are pellets of emerald green with glints of red, which dissolve in water.)

19. Boxed raw lead white. (That is to say, this is boxed in raw form, not altered or mixed with anything else.)

20. Saturated bleached-white. (This is the best bleached-white. It is refined from boxed raw lead white.)

21. Southern black soot. (Southern refers to the southern areas.)

The abbreviated names which the workers who paint in colors use for the various colors are these:

–red ochre is 赭
–cinnabar is called pure red 大紅 or cinnabar red 紅
–red standard is called fat red 膘紅
–indigo is called blue 青 or 藍
–rattan yellow [gamboge] is called moon yellow 月黃
–rouge and carmine are called rouge red 脂紅, carmine 洋紅, deep red 深紅,
  light red 淡紅, peach red 桃紅, or cerise 水紅
–lead white is called white 粉

Furthermore, mineral blue, azurite, malachite, cinnabar, and gold paste are five independent colors that cannot be mixed with other colors, but, with the exception of gold paste, can all be made somewhat lighter with an admixture of white.

Colors can be mixed together, using from two to five colors. The following list is excerpted from Liu Xingmin's color-mixing table, to serve as examples:

–Pink—lead white 1: vermilion 5. (These numbers are mixing proportions and
  should be applied flexibly; so too in the following.)
–Sky blue—lead white 2: Buddha blue 2
–Third blue [light mineral blue]—lead white 1: Buddha blue 3
–Third green [light mineral green]—lead white 1: foreign green 3
–Apricot yellow—orpiment 2: Zhang red [red lead] 5: realgar 1
–Pale grey—lead white 2: Buddha blue 1: black soot (a little)
–Pale purple—lead white 1: Buddha blue 1: saturated red 2
–Pale cream—lead white 2: orpiment 1: Zhang red 1
–Antique bronze—foreign green 1: orpiment 2: Buddha blue 1
–Deep green—foreign green 1: Buddha blue 5: black soot (a little)
–Pale incense color [pale grey]—orpiment 2: saturated red 2: yellow [black?] soot
  (a little)
–Deep blue—Buddha blue 2: Mao blue 2: black soot (a little)
–chestnut color—orpiment 2: saturated red 1: realgar 1: black soot (a little)
–Onion green—Guangming brand red 1: foreign green 2: moon yellow 1: Mao
  blue 1
–Soy color—vermilion 2: Buddha blue 5: Saturated red 3: Mao blue 1: black soot (a
  little)

Besides these, colors such as grass green, light yellow-green, purple, crimson, violet, deep pink, light pink, ochre yellow, ochre green, ochre purple, ochre charcoal, deep grey, fat yellow [pale yellow], fat white [dilute white], sandalwood incense, russet green ["old-leaf color"], ivory, ivory yellow, dark purple, and others, are comparable to the mixed colors used in [traditional] Chinese painting. However, the number of mixed colors used in [modern] colored painting must be greater still.

The artisan painters have original techniques for the application of gold which do not resemble those of the ordinary traditional painter, for whom it suffices merely to dip the brush into gold paste and paint. The gold they use is a foil pounded out of pure gold from Suzhou, called "Suzhou gold." Each book is made from ten sheets and each sheet is 10.7 by 12.7 centimeters. Their methods of application are divided among "gold paste," "gold leaf," and "swept gold." "Gold paste" is the same as for the traditional painters; the gold is mixed with glue to make a fine paste, using a brush dipped into it to draw and to paint. "Gold leaf" uses gold foil applied to a dripped-white ground [a three-dimensionally raised line of white pigment, discussed below] that has almost but not yet quite dried. "Swept gold" is gold powder which is sprinkled on and then brushed smooth and even. As for the amount of gold foil used, these three methods yield a ratio of "one painted [with paste] to three leaf to four swept." Paste is most sparing of gold, while brushing consumes the most gold.[2]

Before there was rubber, pig bladders were employed for the "dripped-white" technique. Alcohol and saltpeter were used to make them very supple. A brass tube was fitted over the mouth of the bladder, the diameter of the tube being approximately one-and-one-third centimeters. Formerly, a copper brush cap [used to cover the end of a Chinese painting brush] was utilized as a dripping tube, allowing three sizes of spout, large, medium, and small, the large spout producing a coarse line and the small spout a fine thread. [The pigment was squeezed out of the bladder, through the spout, and onto the painting surface, like toothpaste from a tube.] Nowadays, we just use a rubber bag to replace the pig bladder, but for the rest we still use brush caps as before. The white used is boxed raw lead white, mixed with Guang glue so that it makes a paste like "almond tea" and packed into a rubber bag. In order to prevent any breaks in the swelling line of white paste, a little bit of cooked tung oil must be added. Then, depending on whether the line of white is to be coarse or fine, cover the spout with a variety of brush cap sizes according to the desired drops or lines to be applied, which are thus called

---

[2]Lu Hongnian, *Wenwu*, 1956.8, p. 16, also discussed three techniques for applying gold: outline gold, which is the equivalent of Yu's gold paste, applied by brush; gilding, or attaching gold foil, which is the equivalent of Yu's gold leaf but not directly associated by Lu with the "dripped white" technique; and spread-out gold, which is done with a needle. Lu gave a recipe for preparing outline gold: use pure gold leaf, place it in a pan with glue, heat it thoroughly and then mix the two manually into a gold paste; wait until the paste turns black in color, then extract the glue and the gold color will be lustrous; combine with glue for use; sometimes a bit of light chicken yellow pigment is added for use, and after it has been applied, in painted gold lines, agate or quartz may be rubbed along those lines, which adds to their luster. In applying gold leaf, Lu specified that the area to be gilt receive an application of tung oil the day before gilding, which is called for also in applying spread-out gold. Lu's instructions for the latter called for gilding a surface with pure gold leaf, then blending gold together with chicken yellow, applying it thickly to the surface, and waiting for it to dry; then, using a needle made of bone or bamboo, spread out the gold in ornamental patterns; Lu noted that this technique was particularly popular in the painting of Chinese figural sculpture.

"dripped white." The method of "dripped-white" paste used in "helmet production" (now used only in the theater) consists in hanging up the spout away from the surface of the painting and then letting the white drip down of its own accord. In this way, the dripping line of white paste can be made to swell up quite high, so this method is called "suspended white."[3]

The Use of Color in New Year's Pictures and Lantern Paintings

On seeing old Chinese New Year's pictures, Fyodor Semyonovich Bogorodskii of the Soviet Union said, "As a type of artistic product, the narrative nature of its content and lively color are bold and very enchanting." (See *Wenyi baocong*, vol. 10, "Record of the Moscow Painters' Symposium.") Further, in Cai Ruohong's article "Creativity in New Year's Pictures Should Carry on the Fine Tradition of Folk New Year's Pictures," he summarizes the masses' "five dos" and "three don'ts" of New Year's pictures. Among the "three don'ts," in "Color that is not bright is a don't," he mentions that "Red should be redder still, green should be greener still." (See *Meishu*, May, 1954). It is not only in New Year's pictures that artists seek for the color to be lively, bold, and fascinating, whether seen from far or near, for the red to be redder and the green to be greener, but also in other kinds of folk painting such as the portraiture and the colored paintings of our previous section where there are the same demands.

1. New Year's pictures

The old woodblock-printed New Year's pictures used colors that were strong, bright, and pretty. They used colors made in China mixed together with colors from abroad. What they cared for were good results—bright and pretty, inexpensive, easy to obtain and easy to use. The colors they used were chosen for their "sharp" and "bright" qualities. "Sharp" and "bright" are art terms that the folk artisans use for color, "sharp" meaning that it stands out, as in the notion of a "redder red, a greener green." "Bright" means intense, so that whether seen from near or far away it is still colorful, shining from within like sunlight, its colors dazzling the eye with their brightness. They demanded that the colors they used be splendid, and so they were bold in their technique and flamboyant, although they still produced paintings in ink monochrome.

2. Lantern paintings

Lanterns are of several types, "palace lanterns," "flower lanterns," and "spring lanterns." Those that are hung up every day are "palace lanterns" and "flower

---

[3]Lu Hongnian's comments on this technique, *Wenwu*, 1956.8, p. 16, closely parallel Yu Feian's. He adds that if the artist wished to conserve on gold leaf, sometimes used to cover dripped white lines, then in making these lines the tube could be pressed close to the painting surface during application, which would produce very low lines, referred to as "scratched white" lines.

lanterns," while those that are seasonal and painted with storybook pictures are the spring lanterns hung for the Lantern Festival [on the 15th day of the first lunar month]. They are well loved by the people. The colors used in painting lanterns are the most delicate and most refined, nearly always using the color standard, which is that portion of a color that is clearest and lightest. With lead white, for example, they only use the standard, which floats on the surface. They very seldom use deep, turbid color. Principally, they stress vegetable colors, but at the same time they also employ foreign colors and foreign "commercial dyes" (such as magenta, emerald green, and aniline blue). The colors they paint with are somewhat more "sharp" and "bright" than those of the traditional painters. The traditional painters' use of color, although beautiful, is delicate and deep and certainly not dazzling. The "sharp" and "bright" that we speak of is showy and gorgeous, not deep but dazzling. For the enjoyment of the masses, they seek to make the lanterns they paint lively and pretty so that when they are set up at a distance they can still be made out. Therefore, in the use of colors, they prefer them somewhat "sharper" and "brighter." At the same time, they still used to be concerned with "distributing their product" (labor handed over its "products" to the merchants), so they had to be all the more lively and pretty, employing whatever catered to the preferences of merchants. However, at the very least, the merchants also had to pick out a few flaws, to make it easy for them to short-change the workers.

Painted lanterns are done in color because they are seen in the daytime and at night. During the daytime only one side can be seen, while at night, when they are lit by a wax candle or an electric light, even the back side is illuminated by the glow of the lamp. This raises the matter of whether or not the colors have been applied with even thickness. Painted lanterns were all made of painting silk (silk gone over with glue and alum sizing). Because the lamplight is placed behind the painted pictures and people see the front of the paintings, if the colors had been applied very thickly or heavily, when seen from the front there would only be a dark shadow and the colors could not be distinguished. If the colors had been applied both thickly and thinly, when seen from the front they would simply be partly black and partly dim, which would not create much delight for the viewers. Therefore, the lantern painters use only colors selected for clearness and lightness. One must realize that the portion of the colors that is clear and light (that is, after they have been mixed with glue and water has been added) is really the essence of the color, which when painted is most bright and beautiful. This is so when seen in daytime, and seeing it with the lamplight coming through at night it is also like this. As for the way they apply colors, the artisans have a unique and ingenious method: in the first place, the colors must be thin so that the lamplight can shine through easily; secondly, they must be even so that the viewer cannot detect so much as a brushmark or any bleeding; if they are completely even and well distributed, the same inside and out, they will shine forth without any dark shadows.

For convenient use in preliminary drafts, the artisan painters have created abbreviations for the colors and abbreviations for directions on methods of mixing colors. Although what is gathered here is incomplete, the more important ones have been selectively listed as follows.

The abbreviations for colors and explanatory notes on painting methods:

1. 紅　　red　　　　　　　　工
2. 朱砂　cinnabar　　　　　朱
3. 朱標　red standard　　　票
4. 銀朱　vermilion　　　　 艮
5. 紫　　purple　　　　　　子
6. 赭石　red ochre　　　　 尹
7. 黃　　yellow　　　　　　卅
8. 花青　indigo　　　　　　主
9. 石青　azurite　　　　　 玉
10. 石綠　malachite　　　　 弖
11. 草綠　grass green　　　 卄
12. 油綠　glossy dark green 由
13. 白粉　white　　　　　　分
14. 墨　　ink　　　　　　　木

The three colors listed above as numbers 5, 11, and 12 are mixed colors, while the rest are unmixed colors. There are also 玉三, which is third blue, and 弖二, which is second green. 木尹 is ink mixed with red ochre, while 尹木 is red ochre with ink mixed in. Furthermore, the art terms and abbreviations for painting techniques are:

丸　wash
火　light wash
通　a complete wash from bottom to top
加　add another
滿　completely washed in and painted
花　this stands for decorative patterns 花紋, striped or spotted patterns 花斑,
　　and variegated dots 花點

So for example, 木丸 is an ink-wash. 木丸加尹 refers to adding a wash of red ochre after a wash of ink. Another example, 火弖 is light malachite, while 弖火 is a light wash of malachite. 通尹丸后加工 means to put down a wash of red ochre from bottom to top and after it dries to add red. 弖卄丸 means laying down a ground of malachite then adding a grass green wash. 票丸艮花 means after putting down a wash of red standard, painting decorative patterns over that with vermilion. This type of abbreviated notation on color methods is unusually simple and convenient.

Chapter 5

The Early Masters'
Application of Colors
and their Methods of
Preparing and Using Colors

With regard to how the early masters applied color, we can judge from the cultural relics, which are fairly abundant from before the time of Zhang Yanyuan of the Tang (this includes both cultural relics handed down and those that have been excavated). However, literature about this from before Zhang Yanyuan's time is extremely difficult to find, and only after Zhang's written records did there gradually come to be more accounts discussing the application of colors. The ways that pigments were prepared and colors were used appeared with increasing frequency in literature after this time.

The Traditional Methods of Applying Colors

In painting, the early masters demanded not only that forms be divided into host and guests [principal and secondary elements] but, furthermore, that in regard to colors their "principals and subordinates be distinguished and that there be harmony among the various colors." The early masters' colored paintings may be classified by types such as blue-and-green, light crimson, ink-wash, outline-and-draw, outline-and-fill, boneless, and others. The painted works from after the Eastern Jin Dynasty provide examples of these types, but the Han tomb murals excavated at Liaoyang and Wangdu do not fully accord with these types. The following examples will serve to explain this.

1. Distinguishing between principals and subordinates and harmonizing colors

The traditional method of applying color was first to visualize the entire surface of the painting and at the same time, for the composition, to calculate in advance which color to use as the principal color and which colors to use for the subordinates—colors which supplement and set off the principal color. This kind of "completed bamboo within the mind"[1] then allows one to harmonize the colors for the entire painting, to coordinate them with one another. For example, some paintings have white as their principal color with other colors as subordinates, such as Dong Yuan's *Xiao and Xiang Rivers* scroll, displayed at the [Beijing] Palace Museum Hall of Painting, which takes several people in white garments as its principal and the rest of the landscape [in ink] as subordinate. Some paintings take red for the principal and other dark colors as subordinate, such as Zhao Ji's *Listening to the Zither* picture (Palace Museum Collection), where the color of the red-clad zither player is the principal and the malachite and other colors are subordinate. Some paintings take light colors as principals and other, brighter colors as subordinates, such as the *Night Entertainment of Han Xizai* in the Palace Museum Collection where, in the fourth section, Han Xizai is depicted listening to the female musicians; there are eight women musicians with patterned robes and skirts, the colors of which are fresh and dazzling and are used as a foil for the light, white clothing, worn by Han Xizai, seated with chest and belly bare on a black chair. This is an unusually striking contrast. The early masters said, "Purple set among blues is the worst of all." They also claimed, "Blue and purple cannot be set side by side." Similarly, they thought that yellow and white used side by side would reduce the brilliance of those colors, so they said, "Yellow and white cannot be juxtaposed." We can see that the early masters paid a great deal of attention to the effects of contrasting and harmonizing colors.

2. Blue-and-green, light crimson, and ink-wash

In landscape painting, in order to show the seasonal aspects of springtime, summer, autumn, and winter, and in order to indicate morning sun, clearing mist, the glow of the setting sun, and so forth, azurite and malachite were used to depict the brilliant green-and-gold brocade of rivers and hills. Some paintings also had cinnabar, orpiment, and white added as adornment for a bright autumn day. Still others employed rouge and white, light yellow-green, and delicate yellow to bring forth the sunlit enchantment of a spring day. Li Sixun's *Bathed in Sunlight* picture used just this method of painting, applying a wash of cinnabar to the ridges and peaks while a covering wash of white was used on the crowns of the peaks, setting them off from the blue-green pines and white clouds below. *Evening Glow* by Yang Sheng of the Tang Dynasty was completely outlined in gold paint.

Wu Daozi, in his plain-line manner, used only light red ochre wash to set off human faces and tree trunks, forming the "Wu style" painting method. The light crimson method consists of ink-wash and light red ochre used together, the tree trunks done in red ochre while the leaves are done in ink, with the light surface of mountains and rocks done in red ochre and the shaded sides of mountains and rocks done in ink. Some used only a light red ochre wash for the tree trunks and human faces, the rest being done entirely in inked texture strokes and wash. Huang Gongwang and Wang Meng of the Yuan Dynasty were the most skilled at these methods.

---

[1] Su Shi wrote: "When a bamboo begins to grow, while it is still but a shoot of one inch the joints and leaves are all there. . . . And so before you begin to paint bamboo you must have the completed bamboo in your mind."

Ink-wash landscapes employed thick and pale ink in place of any color. In some, wet-brush outlines and wash were used, in some, dry-brush texture and rubbing strokes. In some, thick ink was used as the principal and pale ink as subordinate in order to establish three-dimensionality in the painting. In others, unpainted areas were employed as the principal, with thick and pale ink areas as the subordinate to accentuate the spiritual void in the painting. This can be varied in many directions to achieve specific effects.

To sum up, "Skill in applying colors has no fixed methods; skill in mixing colors has no set formulas, only insight into the achieving of a lively application." ("Applying colors" refers to arranging them on the surface of the painting, while "mixing colors" refers to the coordinating of various colors; see Fang Xun's *Shan Jing Ju lun hua.*) We must practice incessantly and acquire experience so that we will be able to achieve a lively use of color. This is true for landscape painting as well as for other types of painting.

3. Outline-and-draw, outline-and-fill, boneless, and their combined uses

Outline consists of using an ink line to delineate the contours of objects, while drawing consists of going over these outlines (ink lines) once again, after they have been covered over by colors. However, the lines to be drawn do not necessarily have to be done with ink; other deep colors can be used for the drawing. For example, rouge used for drawing over azurite increases its brightness, while iron red used to draw over grass green appears to be more realistic, and so forth. With outline-and-fill [*goudian*], the outlines are also first delineated in ink lines, and then, following the inner edges of the ink lines, the appropriate colors are filled in. The relatively opaque colors, such as white, cinnabar, malachite, and others, cannot be allowed to infringe upon the original ink lines, nor can they be separated by any distance from these ink lines. Furthermore, for the filled-in colors, if they are not evenly applied, then contrasts of thick and thin, deep and shallow, dark and pale, bright and dull will occur. The use of color in the outline-and-fill method requires much more proficiency than does the outline-and-draw [*goule*] method. The outline-and-draw and outline-and-fill methods of applying colors were widespread in their use, as can be seen in surviving works of painting from the Eastern Jin through the Northern Song.

The boneless method did not use ink lines to delineate the outlines of objects. In some cases, a composition in ink lines was done in advance on another piece of paper and this composition (draft sketch) was set beneath the paper or silk to be painted on and illuminated from below; then the painting was carried out by using the shadow of the draft sketch as it appeared on the upper side of the paper or silk. There were also some who used willow charcoal to delineate the general outlines of the objects on the paper; then, guided by the charcoal markings, they carried out the painting. Because brush and ink were used to delineate outlines, in antiquity this was explained as the "bone method done with brush," also known as "bone-spirit." The boneless method did not require the use of ink lines to draw outlines, and so it was known as the boneless method. It was not necessarily associated with paintings done entirely in color. Some of these paintings were done in ink, such as the ink bamboo paintings by Su Shi of the Song Dynasty and others. Some employed color and ink together, such as the red plum blossoms by Shen Zhou of the Ming, and others. The idea-sketching school of painting [freely done painting] from the Northern Song on probably evolved from this method.

There was also the combined use of outline-and-fill and boneless. For example, with the chrysanthemums painted by Chen Daofu [Chen Shun] of the Ming, first the flowers were outlined in ink with the brush and then were washed over with some rattan yellow, while for the flower stems and leaves the ink was then drawn out and dotted with the brush to make the forms of the flower stems and leaves. Analyzing this strictly, the flowers were done in the outline-and-fill method—the flower petals were outlined with ink, then filled in with rattan yellow—while the stems and leaves were done in the boneless painting method.

The Early Masters' Selection of Pigments

The early masters' selection of the raw materials for colors has been extensively investigated. As with Zhang Yanyuan's "cinnabar from the wells of Wuling, [cinnabar] sand from Mocuo," and so on, the stress was laid on famous products without concern for their high cost of purchase. This was a different attitude from the folk artisans' strict budgeting and careful calculation, and their selection and use of what was good while inexpensive. This section recounts the important differences in pigments.

1. Cinnabar

As for the selection of cinnabar, aside from Zhang Yanyuan's "cinnabar from the wells of Wuling, [cinnabar] sand from Mocuo," below are the opinions of various artists on the selection of cinnabar for use:

From *Tian gong kai wu (On Using the Products of Nature)* by Song Yingxing of the Ming Dynasty: "Bright, arrowhead, mirror face, and other sands are three times higher in price than mercury, therefore these products are selected and sold for cinnabar."

From *The Mustard Seed Garden Painting Manual* by Wang Gai of the Qing Dynasty: "Cinnabar made from arrowhead is best, while the next best is hibiscus granules."

From *Xiaoshan's Painting Manual* by Zou Yigui of the Qing Dynasty: "Cinnabar from mirror face sand is regarded as superior."

From *Jiezhou's Painting Studies* by Shen Zongqian of the Qing Dynasty: "It does not matter whether the grains of cinnabar are large or small."

From *Painting Trivia* by Ze Lang of the Qing Dynasty: "When selecting cinnabar, it need only be bright and clean, for when it is not clean it is because it has iron mixed in with it and when it is not bright I fear that it is the remains from an alchemist's smelting. Also, there is the kind that has been heated, which is purple and not bright, having turned dark over time. Also available is 'beyond-one-day sulphur' (mercury), the color of which is similarly without spirit. All of these should not be used. One should only select that which is bright red and lustrous."

This writer considers that the essence of what these masters have said is that as long as it is in lump or slab form and has a lustrous surface, then it is good cinnabar.

2. Rouge (that is, cake rouge and cottonball rouge)

Mustard Seed Garden Painting Manual: "Fujian rouge must be used."
Xiaoshan's Painting Manual: "Special quality Hangzhou rouge."
Painting Trivia: "Hangzhou sparrow tongue is the best. What is this sparrow tongue? It is cotton wadded and folded into thin, round pieces of different sizes into which is poured purple stem lac. After it has dried, the trimmed edges of each piece look like a sparrow's tongue, rich and bright in color. None of the others can equal it."

Although cake rouge cannot be found nowadays, still safflower, purple stem lac, and madder can all be found. Because of the color of this kind of rouge, it should be used both in finely detailed and freely done painting. Sometimes carmine cannot replace rouge. (Jiang Sixu sells rouge paste.)

3. Red ochre

Mustard Seed Garden Painting Manual: "Red ochre selected for its firm quality and beautiful color is the best. There is one type that is as hard as iron and one as soft as paste, neither of which should be selected." In addition, "One must select red ochre whose color is bright and glossy and which is neither too firm nor too soft."
Xiaoshan's Painting Manual: "Red ochre with a bright orange color is considered superior, while that with an iron color is inferior. Choose one which has a delicate, fine substance and can be ground up."
Painting Trivia: "The red ochre that painters use today is valued for its firm quality, but that which is hard as iron or soft as paste cannot be used. That with a beautiful color is considered best and that which is dark purple or light like yellow cannot be used."

While a bright-colored red ochre can be selected, sometimes a dark red color is needed.

4. Realgar, orpiment, and yellow ochre

Mustard Seed Garden Painting Manual: "For realgar, select a top-brand brilliant cockscomb yellow." Also, "A realgar selected for being bright and clean can be finely ground." And, "For yellow ochre, use one refined for use over a charcoal fire."
Painting Trivia: "Recently, there is a type of orpiment in Fujian and Guangzhou which comes from the West, not in large lumps but only in fine powder and with no offensive smell."

Orpiment from France is good to use, and its color is delicate and beautiful.

5. Azurite

Mustard Seed Garden Painting Manual: "The only kind of azurite that is suitable for use is that known as plum petal, to which it is similar in form, hence the name."
Jiezhou's Painting Studies: "There are several kinds of azurite, but only the type with a coarse surface and which forms lumps can be used in painting."

Xiaoshan's Painting Manual: "For azurite, select Buddha blue, crumble it up, remove the bits of stone, mix it up fine [in a watery solution], use glue to obtain the standard [the lighter colored pigment that floats to the surface], and this is plum petal."

Painting Trivia: "The azurite produced today includes sky blue, pure blue, Hui blue, and Buddha blue, all of which are different, while Buddha blue is particularly valuable." Also, "There are generally three types of azurite: one is arrowhead blue, one is plum petal, and one is as fine as mustard seed. ... Generally, that which is emerald colored and bright is most valued."

No matter what type of azurite it is, generally if it contains little granulation and is bright and beautiful, it is good to use. Good azurite "stands out" a bit more as well.

6. Malachite

Mustard Seed Garden Painting Manual: "Of malachite, the kind called frog's back is beautiful."

Xiaoshan's Painting Manual: "For malachite, select lion green."

Jiezhou's Painting Studies: "Malachite without much granulation and with a deep emerald green color is fine."

Painting Trivia: "Generally, malachite with a delicate color is good, while that which resembles a frog's back is most valued."

Malachite, no matter whether it is "lion green" or "peacock green," can always yield a delicate pale color, depending entirely upon how it is prepared.

The above are different masters' opinions on the selection of the pigments. Their emphases are largely the same.

The Early Masters' Traditional Methods of Preparing Colors

The early masters' methods of preparing colors[2] were different one from another. By means of considering the actual conditions through a trial manufacture, we can

---

[2]The term which is translated, here and elsewhere, simply as "preparing" is, literally, "grinding and flotation." This refers to two phases of the procedure used for obtaining varied shades of pigments from cinnabar, vermilion, azurite, malachite, and other minerals. The term "levigation" is sometimes used to describe this process, while the most accurate term may be "elutriation." Basically, this process involves grinding the minerals, which will naturally yield varying grades of particles and corresponding shades of color (coarse, medium, and fine particles—the finer the particles, the lighter the pigment). These grades are then separated by liquid flotation (in a mixture of water and glue, in which the finer, lighter particles float to the top while the coarse, heavier particles settle to the bottom) and decantation (in which the heaviest materials are removed from the others by pouring off the lighter materials in suspension above). "First red," for example, a dark red pigment, resents the coarser, heavier particles of cinnabar and is the first to be separated out; once isolated, the remainder is separated by additional stages of flotation and decantation, to produce still finer material.

understand the particular characteristics of each color. Generally speaking, there is only a four-step procedure, "washing, settling, decanting, and depositing." "Washing" means treating the raw material that can be cleansed and washing it in the same manner as rice is washed, after which it is reground. "Settling" means that after the material has been washed and ground fine, liquid glue is added and it is left for a certain amount of time to settle and clarify, so that the clear, light portion floats to the top while the heavy, turbid portion sinks to the bottom. After that, it is "decanted"—that is, the portion that has floated to the top is skimmed off into another dish. The remaining portion that has sunk to the bottom is reground and again "deposited" in the vessel, so that the clear, light color may float upwards and is not forced to the bottom. Undergoing this four-step procedure, cinnabar can be separated into red standard (third red) [light material which floats to the top and produces a pale red],[3] pure red (second red) [medium grade particles which at first are suspended in the middle, and which produce a saturated red], and coarse cinnabar (first red) [coarse particles that settle to the bottom, and which produce a dark red]. Malachite can be separated into green flower, branch green, third green (light green), second green (pure green), and first green (coarse green).[4]

This type of procedure requires a considerable amount of time. The tools used in this preparation include: strainer brushes, mortars, large bowls, large and small dishes, a vent-stove, earthenware pans, porcelain jars, buckets, raw ginger, charcoal, soy paste, and Guang glue. (Bowls and dishes must be fired by first spreading raw ginger juice and fermented soy paste on them and then baking them so that the glaze will not crack when they are heated.) A 300 cc. graduated cup can also be prepared, the better to make out what has floated to the top or sunk to the bottom after the material has been finely ground and the glue has been added.

1. The methods of preparing cinnabar[5]

Jiezhou's Painting Studies: "Xiang You has said, 'Making two hundred grams of cinnabar requires one day of manual labor,' but I feel that it requires two days. However, the more it is ground, the more yellow fat there will be. When it is being ground, a heavy glue solution should be used. After the work is completed, mix it with boiling water in a large cup, mixing it up thoroughly, and leave it to settle for about half a day. Then, pouring out the yellow fat solution, dry it over a charcoal fire. . . . After the yellow fat has been removed, add a pure glue solution and stir it until it is thoroughly mixed. Let it rest for about the space of a meal and then pour it off, repeatedly waiting and then pouring out the yellow fat solution."

Painting Trivia: "Choose [cinnabar] that is bright red and lustrous. Wash it and sun-dry it, crumble it into a mortar, grinding it while dry until it is very fine, and when you are satisfied with it then undertake the process of decanting. After this, use a little bit of glue solution, combining it with warmed river water to make a

---

[3]Red standard is here identified as third red, but elsewhere it is described, and it is generally known, as the yellowish-red portion that floats just above third red.

[4]The pigments are listed here in sequence from finer to coarser particles, from lighter to darker shades. The finer-particulate, lighter-shade pigments are considered most valuable, but all except the very blackish shades have their place in painting.

[5]Vermilion, or synthesized cinnabar, can be prepared in the same manner.

suspension. The decanted material is [still] coarse. After repeated grinding and decanting, the material that is purplish in color is a residue; this residue should be discarded. Now, the first to be decanted is the standard [the portion that floats near the surface]. Floating above the standard is a film. [This is the yellow fat.] Discard this film. Successive decantation will yield three layers. This is generally the same for azurite and malachite. For a large amount, use a bowl. For a little, use a small dish.

"'Third red' [this section of *Painting Trivia* refers to what Yu Feian and most authors refer to as first red]: When [the cinnabar] has been ground up fine, add a glue solution and mix it up evenly, stirring warm water into it. Skim the yellow liquid from the upper part into a bowl, decanting all of the [suspended] cinnabar. The bowl will still contain a coarse residue, which should be mixed up evenly, using your fingers. Again, skim off the yellow liquid into the other bowl—the first bowl. The sunken residue that remains is returned to the mortar and ground again as before [and then deposited]. Now skim off the yellow liquid from the first bowl into a second bowl, and the red which remains at the bottom of the first bowl is known as third red [i.e., Yu Feian's first red, a dark mineral red].

"'Second red': After standing a few minutes [in order to allow for the settling of the heavier materials], the yellow liquid from [the top of] the second bowl is skimmed off into a third bowl. The red which remains at the bottom of the second bowl is called second red.

"'First red' [this section describes what Yu Feian and most authors refer to as third red]: After standing for half a day, the yellow liquid from [the top of] the third bowl is skimmed off into a fourth bowl. The red which remains at the bottom of the third bowl is known as first red [i.e., Yu Feian's third red, a fine-particulate, light red].

"'Yellow fat': There will be a film floating on top of the yellow liquid in the fourth bowl. After this floating film is removed by covering the surface of the liquid with a piece of clean paper, take a dish filled with the yellow liquid and place it on a hand-warmer to dry.

"Cinnabar is divided into three layers. After each decantation, you should use boiling water to remove the glue."

Of the two previously mentioned experts, one [Shen Zongqian, *Jiezhou's Painting Studies*] advocated using a heavy glue for grinding, while the other [Ze Lang, *Painting Trivia*] advocated grinding after washing and drying. If there is a small amount of cinnabar, the former method can be used; if there is a lot, then the latter method is more convenient. However, it is necessary to cleanse the material with boiling water and then grind it. Also, in Ze Lang's method of preparation, including red standard, he created four grades, which involves a rather meticulous procedure. Nevertheless, as for the customary names, what he called "first red" is really third red, while what he called "third red" is generally called first red, and in the same way he inverted the names of first, second, and third for azurite and malachite. In addition, "decanting with water" and "finger-mixing" are procedures for preparing colors, being what Baochai calls "decanting" and "depositing" in the forty-second chapter of *The Dream of the Red Chamber* [by Cao Xueqin]. In this method, after the glue solution is added, according to the difference in the sizes of the color granules, to the ease with which they float or sink, and to the special characteristics of the glue as it floats to the surface when it is heated, one should apply a principle of relative proportions, increasing or decreasing the amount of glue solution in order to differentiate the colors, to establish very clearly each portion, deep or pale.

2. The methods of preparing azurite

Mustard Seed Garden Painting Manual: "For azurite, place it in a mortar, adding a small amount of water and grinding it fine, not using too much force, for using too much force will crush it into a blue powder. Even if you do not use force you will still get some of this powder, but less of it. Just after it has been ground, pour it into a porcelain bowl, add a little clear water, and then mix it up evenly. After it stands for a while, skim off the powder from the surface, which is called the tar. This tar can only be used as powder blue. . . . The middle layer is a good blue. . . . The color on the bottom is very deep. . . . These are called first blue, second blue, and third blue."

Jiezhou's Painting Studies: ". . . Only after grinding it fine, it must be soaked in boiling water and mixed evenly. Let it stand for a while in the bowl, then remove what has floated up to the surface, after which grind it up again."

Painting Trivia: "The method for preparing blue is nearly the same as for red. The sunken residue in the mortar is ground again, glue is added and it is skimmed as before, again dividing it into three layers for use. The more it is ground the bluer it becomes, but do not discard the light portion. All of the ground-up blue must be very finely and lightly ground. . . . When the liquid is skimmed off, it must be skimmed as it is mixed, without waiting a long time. If you wait for a long time, the blue will sink and cannot be extracted. But the floating standard that has been skimmed off into the third dish does not need to be stirred up with the fingers. When it is time to use the azurite, a glue solution must be mixed in, melting it over a fire for use. After it is used, add clear water and dry it over a fire. The glue will float up to the surface and should be completely skimmed off, which is referred to as extracting the glue. If the glue is not entirely extracted, then the next time [the azurite pigment] is used it will be absolutely without luster. So the glue must be completely removed. If you wait and then use it again, you can add a new glue solution at that time."

The methods of preparing azurite, as these three experts have expressed them, lack adequate detail. There is a considerable number of types of raw material for azurite, and when it is being washed and is settling it takes a considerable amount of manual labor. Still, Ze Lang advocates taking the glue out, which is quite necessary.

3. The methods of preparing malachite

The *Bamboo Manual* by Li Kan of the Yuan Dynasty: "When applying color, one must use the best quality malachite. As for method, add a clear glue solution, grind and wash it, and separate it into five grades. Aside from first green, which is too coarse to use, second green and third green can be used for washing on leaves. A lighter color is called 'branch green,' while the extremely pale grade after that is called 'green flower.'. . . If it is to stand overnight, then treat the bowl of green with pure water to take out the glue."

The *Collected Statutes of the Ming:* "From blue and green mineral ores, each half-kilogram yields 570 grams of clean-washed green. Of dark green ore, each half-kilogram yields 1,040 grams of clean-washed green. From one half-kilogram of halogen salt heat-manufactured into halide green, each half-kilogram yields 775 grams."

Painting Trivia: "The method for preparing green is the same as for blue. To use it, add a bit of glue; after using it, take the glue out, which is no different from azurite. The saying is, 'When the green is not green, the glue has not stayed; when the blue is not blue, the glue has not been removed.' This means that azurite is best when the glue has been completely cleaned out of it, while there is no harm in not removing it completely from malachite."

Li Kan divided malachite into five grades: green flower, branch green, third green, second green, and first green, which can be made from the most highly refined, highest quality malachite. The explanation of how each half-kilogram measure of malachite can yield more than that amount, from "The Production of Colored Paintings" in the *Collected Statutes of the Ming*, the proverb that Ze Lang quotes, and Ze's experimentation with not removing the glue from malachite but adding a glue solution when it is ground up to be used again—each of these is entirely feasible. Moreover, these do not impair the color but make it all the more fine and smooth for use.

4. Indigo manufacture

Mustard Seed Garden Painting Manual: "The method for judging indigo is to select for a quality of extreme lightness and for a blue-green which is suffused with red. Then put it through a fine strainer to separate out the bits of vegetable matter. Add a few drops of water with a teaspoon and grind it with a pestle in a mortar until it is fine. If it gets dry, add water, and when it becomes sleek then pound it until it is fine. Every 200 grams of indigo requires one day of human labor to bring out the brilliance of the color. Next, add some clear liquid glue and clean out the mortar by pouring it all into a large bowl, where it is left to settle. Skim off the fine top layer, while the coarse dark layer at the bottom of the bowl should be discarded completely. The portion that is skimmed off should then be placed in the hot sun; it is best to have it dry in the sun for a full day. If it is left to the next day, the glue will become lodged in it. Every other color can be made in all four seasons; only indigo must wait until the hot season. In painting, there are a great many occasions for using this color, for its color is the finest. Other colors can all be prepared in a single day, while only indigo requires several days; for other colors, the four seasons are all suitable, while for indigo, only a summer day is suitable for adding glue, preparing, and removing the dregs, since it should be dried in the hot sun and not by the heat of a fire. If it is urgently necessary, then dry it by a fire, but only if it is not allowed to become dried up or scorched will it be good."

Xiaoshan's Painting Manual: "For indigo, using Guang blue with a faint grape color is fine. (This is directly counter to Wang Gai's [selection of an indigo] 'which is suffused with red,' for 'indigo' and 'Guang blue' are not the same thing and 'Guang blue' is not suffused with red [Wang really meant] indigo which is expensive.) Put it through a strainer to remove any residue, use glue while grinding it up fine, then wash it to get the standard, which is then poured into a dish and dried over a gentle heat. In the summer months, it will dry faster if divided into dishes, but the glue may smell foul."

Jiezhou's Painting Studies: "Flower blue or indigo. . . . Its color is blue-green and vibrant and it is an essential color for painters. First, crush it into a paste (this refers to the kind that has been mixed with lime to form lumps) and soak it in boiling water. At the first steeping, a yellow liquid will come forth. What comes out forms a film, but although it is steeped several times, its original color will still remain fixed securely by the lime. After it has been ground fine in a mortar, pour in some glue and mix it evenly in a large bowl, then wait for an hour or so and pour the color that has floated up to the top into another bowl. Then, taking what has become anchored (meaning that which has sunk and become attached to the bottom), and not necessarily adding glue, grind it up fine as before and then pour in the color that had previous floated up. Waiting for an hour or so, pour this into another bowl and continue in this manner several more times until the color on the bottom becomes somewhat pale. Since indigo is made with the addition of lime, the color that emerges gets successively better. Moreover, it cannot come forth completely in one or two turns, so it must be done numerous times to attain it. ... Combine the liquids that have been poured off and wait altogether for a half day or so, then pour them into a porcelain dish and again remove the portion that has become anchored. Place the porcelain dish over a charcoal fire smothered with lime-ash and, while it is drying, use something to mix it up evenly with care. If you perceive that it is drying without becoming finely mixed, then the upper half has most of the glue and the lower half has most of the lime. It must be stirred during the time it is drying or the lime will not be completely mixed together with the viscous glue."

From *Models for Landscape Painting* by Fei Hanyuan (published in 1792): "The method for pounding up indigo is first to put the indigo through a strainer to filter out the lime and vegetation, until it is clean, then add a little bit of glue and using a mallet made of rotten wood pound it fine, giving it a thousand strokes before turning it. Then add a little more glue and pound it again. Do this several times until you see no more residue, then pour in some clear water, not too much and not too little. Pound it again. Wait until the liquid settles out clearly, then remove it, and then pour the indigo into a clean utensil and dry it in the sun. If there is no sunshine, it can be used after drying it over a low fire."

Painting Trivia: "The method for preparing indigo has been very generally discussed in recent painting manuals. ... Indigo is derived from a vegetable material that is extremely light and supple. If you grind it in a mortar, the hard substance cannot overcome the soft, therefore make it into a paste by hand, for the soft can overcome the soft so that the residue is completely blended into a thick liquid. Manufactured indigo is used as a paste. Begin by using silk as a strainer and straining out any bits of vegetable matter. Make a thick glue, approximately two hundred grams of indigo to one hundred grams of glue, then grind it up, mixing and shaping it into pellets like small pills and sticking them to the bottom of a large porcelain dish; they cannot be dried in the sun nor can they be dried by fire, but wait until they have dried by themselves. After that, steep them in clarified river water for a day and a night and a yellow liquid will emerge. Each morning, skim off the yellow liquid and replace it with clear water. After ten days or more, the yellow will not be exhausted but the glue will be completely gone. Heat it until it is dry again and mix it with glue, soaking it in water as before. After another ten days or more, using the complete emergence of the yellow liquid as a guide, dry it over heat and gather it up to await grinding it up for use. ... Over a hot hand-warmer, make a large dish of some very thick glue, put four or five drops into an empty dish, add a little bit of indigo and make it into a fine paste with the fingers, just as in the method for making gold paste. When the paste is nearly dry, dip your fingers into the water and rub the paste until it is extremely fine, gleaming and brilliant, and then add several drops of clear water to the paste to soften it, not too much water and not too little. If there is too little the glue will be too thick, while if there is too much it will be too watery. One should be careful in doing this, but too much is better than too little. After the paste has been softened, put it back into a large bowl to save, but if the bottom of the dish is moist, slippery, and not sticky to the touch, you must wait for it to dry by heating it before using it again. Or select another dish, making a paste like before, and substitute this one for the other. If four or five people can be assembled to make paste, two large bowls can be produced in a single day. When the large bowl is full, the top must be covered. It should settle overnight; in the early morning you should use a piece of thin, unsized paper to take off the film that has floated to the top of the bowl, then carefully skim off the blue liquid and store it in another large bowl without drawing off any of the sediment. Afterwards, place a three-inch dish containing part of the blue liquid over a blazing fire to dry, but in the meantime do not add any cold water. When it is drying, wait until it is just dry to take it off, and do not let it scorch. Wait until it has cooled, then put the dish in a damp place for about half a day, allowing it a little moist air. Carve it into pellets, perhaps wrapping the leftover fragments in paper to store for later use. When it is to be used, place it in a porcelain dish and add some drops of clear water, preparing it as the drops are added; each time it is used, it will be absolutely flawless. This way of making indigo is a hundred times superior to using a mortar. ... When making a paste of indigo, exercise discretion with the glue solution, for the glue should be light, not heavy."

In refining indigo pigments from indigo, Wang Gai and Zou Yigui said to use "Guang blue" or "Guang indigo." Neither of these two used lime-soaked pigment. The three masters Shen Zongqian, Fei Hanyuan, and Ze Lang said to use lime-soaked lumps of indigo. As for their statements about "crushing it fine," "grinding it fine," and "finely ground," these all ought to be "pound." Indigo in lump form should first be ground and then "pounded." I consider indigo to be one of the most important colors in Chinese painting. Therefore, the different masters' methods for refining indigo pigment from "Guang indigo" (in powder form) and "indigo" (in lump form) have been written down here in particular for use as reference material.

5. The preparation of other colors

The methods for preparing the various colors discussed above are rather complicated; for all the other mineral pigments, the above methods can be used to separate them into their various different shades of deep and pale. For example:

Realgar. For the powdered form, first boil it in water, dry it in the sun, then add heated alcohol, grind it up fine and add glue before preparing it. For the lump form, which comes in both deep and pale tones, the deep shades are prepared together, separately from the pale shades.

Orpiment and yellow ochre. These should also be boiled before preparing. "French orpiment" has no foul smell and does not need to be boiled. When prepared it will yield three tones of deep and pale.

Red ochre. This should be boiled first. It has different shades of deep and pale when prepared.

Earth red and white chalk. Do not boil these. Only begin the preparation process, retaining only the uppermost layer for use.

Lead white. The masters have set forth a great many methods for preventing "reverted lead" [lead white which has turned black over time]. However, none of them is very effective.[6] Nowadays, there are titanium white, zinc white, and clamshell white, which can be substituted for lead white.

As for the vegetable colors, rattan yellow is best when used with a brush dipped in water. Cotton rouge is decocted in water and the juice wrung out; then it can be used. *Sophora japonica* blossoms are picked, scalded in boiling water, molded into cakes and dried in the sun as preparation for use. Gardenia yellow and logwood are decocted as they are used.

The Early Masters' Methods of Using Colors

Each of the ancient masters had a different technique for using colors, so they cannot be discussed in generalities. Moreover, the methods for using colors of the painters of the Sui, Tang, Five Dynasties and the two Song periods are scarcely recorded in literature. In some cases there are merely "a few words or single characters." What is written herein, although entirely ordinary in terms of methods, is especially important as it can establish a link with the particularly fine tradition of painting in color of the two Song dynasties.

1. Li Kan on "underlaid washes" and "overlaid washes"

In the book *Bamboo Manual,* he wrote, "'Underlaid washes' are a most important aspect, for you must distinguish shallow and deep, back and front (the back or front of a leaf), dark and light. When using a wash-brush to break [the tones, creating modulated layers of tone], avoid the appearance of brushmarks, and if you want [the bamboo] to seem to grow as a single unit this depends entirely on your success in wielding the paint brush. (This means to use two brushes, one dipped in

---

[6]One such method is that of Zhou Jiazhou (d. ca. 1660), recorded in his *Zhuanghuang zhi,* or *Book of Mounting:* "Melt a lump of pure white soda in water, just as if you were going to wash clothes. Then dip a new brush in this solution and therewith dab the spots that have turned black, taking good care that the soda solution does not spread over the edges of the spots treated. Then take a sheet of *lien-ch'i* [lianqi] paper, and having spread that over the scroll, roll them up together. If after the lapse of two weeks you take away the covering paper, you will find that all the black spots have been transferred to the sheet of *lien-ch'i* paper. If the dirt should not have disappeared entirely, you should treat the painting in the same way for a second time. If the dirt consisted of only a thin layer, it will disappear when the painting is treated once or twice. Paintings covered with a layer of many years will become white again after having been treated three or four times. When the white pigments are again as bright as if they were newly laid on, you should finally dab the spots treated with thin tea water newly brewed, in order to remove what is left of the soda." Translated by R. H. van Gulik, *Chinese Pictorial Art as Viewed by the Connoisseur* (Rome: Instituto Italiano per il Medio ed Estremo Oriente, 1958), p. 302.

color, one dipped in water, first painting the areas that should be dark in color, then using a wash-brush that will modulate and break [the tones], making them progressively lighter as you go. However, it works as well to use a single brush, dipping it in water first and then dipping the brushtip into color to pick up the wash.) If you are not especially careful there may be some puddling, which will ruin all of your previous efforts. In this method, 'Formosan indigo' (which resembles Guang blue but is not suffused with a grape-red color, and which is produced in the South Sea islands and sold at Chinese pharmacies) or Fujian 'snail-shell blue' (indigo) is placed in a bowl. ... According to the veins in the leaves (this means, for a bamboo leaf, according to which end is the tip of the leaf and which end is the base), go down the center with a dipped brush for an underlayer of wash (brush the wash along the center of the leaf). New leaves take a light wash, while old leaves take a concentrated wash. Along the branches and stalks, the recessed areas take concentrated washes while the shallow areas take a pale wash, all according to their relative tonality. (In this section, after the forms are outlined and before green washes are applied, indigo is first used for a ground that varies from dark to light, which is the first step in applying washes in green. In both Tang and Song, the method for all leaves done in wash was the same as this.)

"The method for mixing green (malachite) is first to mix in thick glue and grind it evenly, then add decocted sophora yellow and mix it according to the lightness or darkness to be obtained. ('To be obtained' means that which is suitable or fitting. Mixing with sophora yellow is a method handed down from Tang and Song. Sophora yellow has taken the place of the juice of 'Sagina maxima'; see Zhang Yanyuan's discussion of color.[7])

"The method of moistening the brush should allow it to rub on a thin layer; it should not be heavy or thick or leave traces of the brush. Equally, when laying down lines of ink that come to an intersection, do not let them protrude or fall short unevenly and especially do not allow any white to show. ('Showing white' refers to white paper appearing between the ink line and the color [wash]. This is the outline-and-fill method, where the color is filled in within the ink lines, not the outline-and-draw method, where the color covers over the ink lines and then the lines are drawn over again.)

"'Overlaid washes' are what tie the painting together, so they must be especially meticulous.[8] Wait for the applied color to dry, then inspect it carefully for any places where there are gaps or omissions, take a clean, dry cloth and, with some exertion, rub it wherever you fear the color has been omitted, and this can conveniently restore it to an even application. Except for the backs of leaves (the backs of leaves are light-colored) everything uses 'vegetation juice' for the overlaid washes ('overlaid wash' means a covering layer of wash). The backs of leaves only use a light rattan yellow for overlaid washes. ('Vegetation juice' is grass green, also known as 'juice green,' a green color made by combining indigo and rattan yellow.)"

I have experimented with these methods of Li Kan for underlaid and overlaid

---

[7]See Chapter 1, note B, p. 19.

[8]"Overlaid wash" might also be translated as a "basket wash," a "snare" or "loop wash," an "enveloping wash" or "casing wash," as it is sometimes modulated to produce a darker or lighter tone around the perimeter of an object. For further discussion of this technique, see below, pp. 75ff.

washes and found that not only bamboo leaves but any kind of leaves can be done by using this method. No matter if it is on silk or paper (sized paper), first put down dense and light washes of indigo to differentiate the reverse and main sides, front and back, *yin* and *yang,* and their relative brightness in terms of light and shade. After the indigo washes are applied, use second green or third green in a very light wash over that. In the places where the indigo is heavy, the malachite must be somewhat thinner, while in the places where the indigo is thin—the places where it is brightly lit—the malachite should be applied a little more thickly. In this way, in the places where the indigo appears heavy, the malachite looks deep and dark, while in the places where the indigo is thin or absent, the malachite will appear bright, this being the upper face of the leaves. The backs of leaves are generally a little lighter than the fronts, or a little paler, and for this a wash of "green flower" is used. A wash of "branch green" is used on the delicate branches. When you want to make a wash of malachite, glue is first added to the malachite, then sophora yellow is added and mixed. Done in this way, not only will it increase the freshness and brightness of the color but also will increase its adhesiveness. After it has dried, wipe it with a cloth to see if the malachite is firmly attached. If you lose some of the color, then do not simply repair those areas where the color was lost but go over the entire area that was washed in malachite with a very light layer of sophora yellow mixed with glue, applying it with a soft goat's-wool brush. Wait until the malachite color is firmly set, then add an overlayer of grass green wash composed of rattan yellow mixed with indigo. If one layer is insufficient yet another one can be applied. The backside still must be painted.[9] This is the principal method for using malachite and adding overlayers of wash in our fine national tradition of painting.

2. Wang Gai and Ze Lang on the application of white

Wang Gai wrote, "Whether painting white flowers or ones of mixed colors, whenever it is done on silk with white pigment used on the front, the backside must be painted. . . . The method for applying white: white should be applied to the front, thin and lightly, up to the ink outline but not over it or short of it. If one layer is not even then add another layer, so layers should be thin to facilitate adding on.

"The method of doing washes in white: for peony and lotus, even when a [color] wash has already been applied, a wash of white must be added to their tips so there will be layers of deep and pale. If you wish to make flower petals delicate and beautiful, then you must first build up washes on a ground of white.

"The method for fine lines in white: for flowers such as lotus and autumn mallow, the veins of the petals should be drawn in white; in tinted chrysanthemum flowers, each petal also has long veins which should be drawn in detail with white as well as outlining their outer form, after which the colored washes are added.

"The method for dotting in white: flowers in the sketched-from-life manner are not outlined but only given darker and lighter dots by a brush using white dipped in color. . . . When the flower pistils and stamens are dotted in with white, they

---

[9]This painting of the backside [literally, "accenting" or "lining" the backside] refers to the technique of adding certain appropriate colors to the backside of the silk in correspondence with the forms and colors painted on the front, in order to bring out their color values.

must be mixed with rattan yellow, although they cannot be made too dark, and the glue that is added should be light. When dotting in the yellow pistils and stamens, they should be raised along the perimeter with a round depression on the inside and they should not be too dark.

"The method of painting the backside with white: for all white flowers done on silk, the backside must be painted with thick white to make them stand out. Even if the front is done with various light colors, the backside can only be painted in white; if they are done in heavy colors, so you feel they will not stand out, then they should still be painted on the backside by means of white mixed with these colors. For the backs of leaves, if a light green is used on the front, the backside can only be painted by means of a greenish-white and not with malachite."

Ze Lang wrote, "In the method of building up white, for flowers such as peonies or lotus on plain silk, to cover over the draft sketch use white for a wash that conforms to the petals, with a thick white for edges of each petal and another brush dipped in water that is washed on toward the base—this is called built-up white. After it is built up, apply each color in a wash from the base outward, leaving the edges white. After the washes are completed, it really seems as if each petal is suspended in space, ready to be moved by the wind; this is fine-detailed painting technique of the highest order. To make dewdrops, make dots with a brush dipped in a thick white; while drying, there will be a depression in the center of each dot and there is no harm in that. By extension, to paint large red peonies, cinnabar should also be built up as in the method for building up white. After applying a wash of rouge or carmine from the petal's edge to its base, paint the backside with light red standard, which will make the red bright and deep."

Looking at these two masters' methods of applying white, they both advocate a two-sided application of white (painting the backside). In using clamshell white, it is only suitable for use on the front side. For painting the backside, it is best to use titanium white or zinc white.

3. Various masters on the application of cinnabar

Wang Gai: "Red standard is used to paint people's clothing. Good cinnabar is used to paint maple leaves, railings and balustrades, Buddhist and Daoist temples, and other things. The bright middle layer is dried in the sun, combined with glue, and used for the large red petals of camelias and pomegranates, together with a layered wash of rouge. The heavy material that sinks to the bottom can only be used for painting the backside."

Shen Zongqian: "The yellow fat liquid is poured off. . . [Then, the cinnabar] is used for coloring human flesh and, mixed with all kinds of yellows, for clothing, being much more brightly colored than red ochre. After removing the yellow fat . . . [what remains] can be used for finely rendered human clothing and for all types of red leaves that dot the landscape."

Ze Lang: "For washes of crimson, take second red as the ground layer, use a wash of light rouge six or seven times, then use thick rouge for a fine outline and it will be naturally bright and beautiful. If the modulated washes are sufficient and a layer or two of alum is applied, then even after the paper or silk has been worn out the color will remain bright, which is what is known as protecting natural purity by means of human effort. There is also a little redder color that can be extracted from beneath the yellow fat which, added to second red, will serve as the ground. If at first it seems to have a yellow color, wash over it with carmine as many as six or seven times until it is extremely red and afterwards outline it with rouge. If the silk is painted on the backside with 'third red' (this should be first red) or with lead white, the red will seem twice as bright. . . . If rouge is applied in several washes, it will become dark and blacken, whereas several washes of carmine become deep and remain bright. When washes are added repeatedly over cinnabar, it becomes doubly beautiful and attractive.''

From what these three masters have said about the application of crimson color, one should not only use second red but should also use rouge and carmine as layered washes, as well as using heavy glue, covering them with alum, and painting the backside. Ze Lang advocated using carmine for layered washes, which increases the brightness of the color more than using rouge, which is precisely correct.

4. *Various masters on the application of blue and green*

Wang Gai: "Whenever blue and green are used on the front side, the backside must be painted with blue and green, and the color will then become full. . . . If the light, pure-colored, pale upper layer (of azurite) is used as a wash over green leaves on the front side, then you can achieve a deep, rich color. The middle layer has the suitable degree of fineness and the appropriate shade to be used for pure blue flower petals and the backs and heads of birds. The heavy, deep-colored layer on the bottom is used for the wings and tails of birds and for painting the backside behind deep green leaves. When painting the bodies of birds and petals of flowers, for [painting the backside behind] pale blues use layered washes of indigo, while for [painting the backside behind] deep blue use layered washes of rouge. The upper [*sic*] deep-colored layer (of malachite) is only suitable for painting the backside behind dense, rich green leaves and green, grassy ground and slopes. The middle, slightly lighter layer is suitable for painting the backside behind grass, flowers, and green leaves, or is used on the front covered by grass green or in painting kingfishers using a faint grass green wash. Its lower [*sic*], lightest color is suitable for doing the reverse side of leaves. Whenever malachite is used on the front, always use grass green for outline and wash. For depth, grass green should include blue, while for lightness grass green should include yellow. (Author's note: Wang Gai said of malachite that 'its upper layer is deep-colored, its lower layer is lightest,' but it should be that the top color is the lightest and the bottom is the deepest.) On silk, if grass green is used on the front, only malachite is suitable for painting the backside; if dense, heavy colors are used on fans or paper, they cannot be painted on the backside, so when using them on the front apply grass green as an overwash and then they will still appear deep and full. Do not pile them on thickly in one application, but gradually add several layers and then the color will be even and without brushmarks."

Shen Zongqian: "On all mountains and rocks: for those mostly of blue, use malachite for embedding the moss; for those that are mostly green, use azurite combined with malachite to embed the moss. If the brushwork is meant to be sparse and loose, then the application of color ought to be commensurately light. When using blue and green mixed together to cover the mountains and rocks, then only light malachite is used to lay down the grassy ground and slopes, and moss need not be applied."

Ze Lang: "As for the use of green, it should be of several layers added successively and cannot be piled up thickly in one application. When using green on the front side of paper, it should be covered with a grass green. If grass green is applied on the front of silk, then the backside is painted with malachite, which can only be used lightly and not spread on so thickly as to alter the original grass green color, for that would reverse the feeling and decrease its interest. Second blue can be used for fringed iris and as a wash over lotus leaves done on the front side in dark green or various other colors. (Third blue) can be used for flowers such as morning glories and 'azure-eyebrows' and for dotting in mixed leaves and various grasses, as well as for people's outer clothing and for painting the backside of silk. For deep blues, use rouge for outlines and washes; for light blues, use indigo for outlines and washes. In earlier times people used ink outlines for the folds of clothing, while flowers and grasses were outlined with purple; ancient paintings allow for the subtlest appreciation." (Author's note: Ze Lang reversed the first blue and third blue of azurite. He also followed Li Kan in the use of malachite, which he said must be mixed with sophora yellow, omitted here).

What these three masters have said, above, about the methods of applying blue and green is: first, paper and silk are not the same; second, they all said to paint on "painting silk" (silk with the threads pounded flat, sized with glue and alum) and not to paint on "unfinished silk" (raw silk, silk with round threads that have not been pounded flat, with no glue or alum added). In the use of blue-and-green during the Tang and Song periods, because artists used "unfinished silk," whatever was painted on the front was also painted on the backside and a covering wash of "grass color" was added to the front. Contrary to what Wang Gai and others have said, grass green was used on the front and the backside was painted with malachite—however, they also said to use azurite and malachite on the front side. (Li Kan also said to use "unfinished silk.") In addition, before applying blue or green, some of the painters of the Tang and Song first used indigo for layered washes to produce deep and shallow, back and front, and dark and light ([in conjunction with] malachite), while others first used layered ink washes for deep and shallow, back and front, dark and light ([in conjunction with] azurite), after which the blue or green was laid on in layered washes. These three masters limited themselves to applications on "painting silk." However, if you use painting silk, and if you use the methods of these three masters, the result will be unusually fresh and beautiful.

5. *Various masters on the mixing of colors*

These are the different masters' methods for combining colors, leaving out the *Technique of Painting in Colors* of Wang Yi of the Yuan Dynasty.[10] I have gathered them together and entered them into a single table. Although what these masters said about the mixing of colors is limited, we ought to make vital use of it. to be sure, the blending of colors does not end with this.

[10]Wang Yi's *Technique of Painting in Colors* is translated above, pp. 32ff.

Mixed colors	Pure colors									
	Indigo	Rattan yellow	Red ochre	Ink	Rouge	Carmine	Red standard	Cinnabar	White
Grass green	5	5							
Deep green	6	4							
Light green	3	7							
Bud green	2	8							
Glossy green	5	4		1					
Dark green	4	5	1						
Lotus blue	2				4	4			
Lotus root	2				3	3			
Golden red		4					6		
Flesh color			3		3				4
Silver-red					3		3		4
Dark red					4			6	
Pink						4			6
Golden yellow		5			5				
Dark yellow		4	6						
Deep red			4					6	
Dark purple	3				2	5			
Iron color		7		3					
Soy color		6		2	2				
Sandalwood incense		5	5						
Autumn incense color		8	2						
Gosling yellow		8					2		

[Figures indicate proportions in tenths. For example, grass green is mixed from two pure colors, 50% indigo and 50% rattan yellow.]

Chapter 6

Modern Traditional-Style Painters' Methods of Preparing and Using Colors

Methods of Preparing Colors

Some methods used by modern traditional-style artists for washing, settling, decanting, and depositing colors are not very consistent with ancient methods, and some are more precise than the ancient methods. In order to save time, some people have even begun to use a small-scale pulveriser to grind up the pigments.

1. Cinnabar

Take raw mirror-face cinnabar and grind it, dry, in a mortar; the finer it is ground, the better it is. When you have 200 grams of cinnabar ground up, then put the finely ground cinnabar into a bamboo tube with a diameter of six centimeters or more. The bamboo tube should have the base remaining at the lower joint, and be washed and bound with lead wire to prevent it from splitting. Also decoct some Guang glue—oxhide glue solution—to a thick state and, taking the clear, light liquid on the surface, pour it into the bamboo tube. Mix it very thoroughly with the cinnabar, adding clear water as it is mixed and let it rest for an hour before use. Fill an earthenware pan more than half full with water, then place the bamboo tube horizontally [upright] in the pan and heat it over a low fire. The water should not boil, so add cold water as it heats. Finally, heating it until the cinnabar in the bamboo tube has nearly dried out, remove the pan, holding it level, then wait until the water cools and take out the bamboo tube. Now wait until the cinnabar in the bamboo tube is completely dry and undo the lead wire around the tube. Do not let the tube split apart of itself, but use a knife to gently cut it into two halves so it is ready for use. At this point, the cinnabar in the tube has an upper layer of red standard which is increasingly yellow toward the top, a lower layer of first red which is increasingly purple toward the bottom, and a middle layer of pure red, second red, which is particularly fresh and bright. This method saves time and work over the "liquid decantation" method. However, three supporting feet must be sawed into the bottom of the bamboo tube [below the base segment] so that the hot water inside the pan can flow through it. The best heating stove that can be used has a charcoal fire while the next best is to use an oil-fed vent-stove, because it is easy to regulate the fire in both of them.

Using a knife to open the bamboo tube, place each of the three sections, which have been decocted into red standard, second red, and first red, into a large bowl, soaking them in boiled water and steeping them for three or four hours. Pour off the water, mix them evenly by hand, then again soak them in boiled water and steep them for several hours. Wait for them to settle and clarify; red standard may require more than twenty-four hours. By this time, the glue has completely floated up. Skim off the liquid, dry the pigments in the sun or by heat, and place them in a moisture-proof place, preparing them for the time when glue is added for use, which is called "extracting the glue." With azurite, after it is used the remaining portion still must be soaked in boiling water to "extract the glue," while after using cinnabar it is not necessary to remove the glue again.

2. Red ochre

Red ochre in lump form can be bought at Chinese pharmacies. Dry-ground to a very fine degree, it can be employed in the same bamboo-tube preparation method set forth above; however, in the lowest layer of heavy color, after it has dried a magnet should be used to draw off the iron remains within it, then apply this preparation process once again. The upper layer will be suffused with a yellow color; the middle layer will be the basic color of red ochre. The lower layer, already suffused with the dark red color of the iron that has been drawn off, will be just what the early masters called "iron red" color, somewhat more red than red ochre mixed with ink and somewhat darker than rouge; it is always used in painting figures and utensils, in painting the veins that are drawn on the surface of flower stems and leaves, and in painting the fur and feathers of sparrows, wild ducks, wild geese, eagles, sparrow hawks, horses, camels, and other creatures. "Iron red" is an important color used by the ancient masters and by folk artisans. Painters from the Ming and Qing on have abandoned its use, using only red ochre mixed with ink as a replacement. The painters of the late Qing–early Republican period only used the "standard of red ochre" from the upper surface of red ochre.

3. Carmine or "Western red"

Carmine is a product of old Germany, a product manufactured by the Greater German Pigment Company and by the Pelikan Company, the former being a powder and the latter being in lump form, with the lump form being called "Western brick red" in China. We know only that it is a pigment made from animal remains [from the *Dactylopius coccus* or cochineal insect]. Ze Lang has already described this in his *Painting Trivia*. Its advantage is that when painted on sized paper or on silk it absolutely will not corrode through to the back side; if a white goat's-wool brush is dipped in it for painting, it will not stain the brush hairs red and they will remain white. Other kinds of carmine generally will corrode through to the back side (folk artisans call this corrosion "biting") and will also stain the brush red. When we use them, it is only with the addition of glue and when ground up fine (the fingers are used to grind it up) that they become usable. They are particularly susceptible to moisture. After use, the remaining pigment must first be dried in the sun, then covered and stored. Otherwise, the color will become dark and dull, not bright.

4. Sophora yellow

Sophora japonica flowers are found everywhere in the north. When they are being picked and gathered, the first necessity is to select the open blossoms, not the receptacles but only the flower petals. The second is to get the unopened buds, plucking them off together with the connected receptacles. The two types should not be mixed together but should be boiled separately in hot water, molded into cakes, and dried in the sun for use at the appropriate time. In order to enhance the luster of malachite, "green flower" and "branch green" are used mixed with a decoction of *Sophora japonica* flower petals, while "second green" and "third green" are used with a mixture of the two types, decocted *Sophora japonica* flower petals and *Sophora japonica* flower buds. The proportions for mixing the decoctions is one to two. Blue-and-green landscapes use this method, as do paintings of flowers-and-plants and feathers-and-fur.

The use of sophora yellow mixed with malachite is mentioned in the *Bamboo Manual* by Li Kan of the Yuan. It not only enhances the beauty of the color of the malachite but is also used to create a fused effect; at the same time, grass green can still be used as an overwash. Painters from the Ming Dynasty on seldom used sophora yellow mixed with malachite. In paintings of these two dynasties, when malachite was used on the front an overwash of grass green was very rarely added.

5. Azurite

Azurite in lump form with a coarse exterior is called "Hui blue." There is a flat slab form clearly demarcated into different tones, which is called "Yunnan blue." There is a type resembling small grains of rice (or cereal grains) of varying size with glittering particles on the surface, called "granulated blue." There is one type with different sized lumps with a surface of gleaming kingfisher-blue, called "Tibetan blue." There is a kind made into a powder which is mixed with mud and called "paste blue." The former four types, before they are ground, are broken up by boiling them in an earthenware pot for an hour. There will be a muddy froth floating on the surface which should be skimmed off with a ladle, skimming it as it boils. After boiling, wait until it is dry and then grind it up. However, for this kind of earthenware pot, one that is manufactured in the south with a soy-colored ceramic glaze on the inside must be used. The insides of the northern so-called "Lishan thick-earthenware pots" are coarse and unglazed and absolutely unsuitable for use. Of these raw materials, the most widely used is paste blue. Its color is dark and dull since it is mixed with mud and it requires a great deal of labor to make it up.

With the first four types mentioned above ["Hui," "Yunnan," "granulated," and "Tibetan" blue], after they have been ground up fine in a mortar, put into a pot, and boiled again, there will still float up a muddy froth which should be skimmed off with a ladle as it floats up, until there is nothing more on the surface. At this point, the color will already be sufficiently bright. Wait for it to cool and settle to the bottom, then add a clear glue solution and stir it vigorously. Stir it until a pale greyish-blue color floats up. Using a bamboo tube according to the method for preparing cinnabar will bring out the four different shades of "powder blue," "third blue," "second blue," and "first blue." As for the latter type, "paste blue," the principal method is first to boil it, stirring and skimming it as it boils until the floating mud is completely skimmed off of the surface. After it has cooled and settled to the bottom, dry and grind it. After it is ground up fine, boil and skim it as before, boiling it until there is no more floating mud and a bright blue color floats up. Let it settle out, then add glue and use the liquid decantation method—the method of Wang Gai, Ze Lang, and others—to produce the three colors. Because it undergoes the addition of glue and some of the muddy froth will float up, which is not suitable for the natural floating and sinking in the use of the bamboo tube, the flotation must be done by hand, controlling the decantation and depositing.

As for calculating the proportion of color produced which is of "outstanding" grade, "Hui blue" and "Yunnan blue" can yield as high as sixty percent good blue (including second blue and third blue); "granulated blue" and "Tibetan blue" can yield up to fifty or fifty-five percent; "paste blue" yields thirty-five percent at most, and some yields only thirty percent. The method for removing glue from azurite is the same as for cinnabar. But azurite requires more thoroughness, so it is best to soak it once in boiling water so that there is no remaining glue. Glue is added to suspend the blue, and this should be somewhat thicker than the glue added to the cinnabar and somewhat more in quantity. Azurite that has had the glue removed, when stored away, as long as it is free of moisture, will still preserve its brightly colored beauty no matter how many years it is kept.

In addition, when buying already prepared azurite (Jiang Sixu has paper packets of five grams, and in Beijing there are cakes of five-gram and ten-gram weights), the first step is to grind it up in a mortar and the second is to add clear glue, mix it evenly, and check whether or not any mud floats up; if not, it can be used immediately, otherwise the preparation process must be applied once more. If the colors still are not clear when prepared, then it is necessary to boil them and to make sure that the mud completely floats up. In this way, although quite a lot is lost you can still obtain a good blue.

The five colored inks of the Qianlong and Jiaqing eras, which we regard as good raw materials of that time,[1] when prepared again can yield very good colors. The white ink of the five colored inks is manufactured from mother-of-pearl that is produced in Gansu and Xinjiang. This type of white ink must be steamed in a steamer until it is dissolved, then the glue that has floated to the top should be removed and finely ground in a mortar, and then it will be much better for use.

6. Malachite

There is a malachite in lump form with a coarse surface and entirely of one color, not differentiated into dark or light. There is a lump form with a surface sectioned into round raised and lowered bands of varying tones. There is a kind that has a black linear pattern mixed into it. One kind has kingfisher-green colored patterns that resemble peacock plumes. One is crushed fine into grains. One is naturally divided into small, thin slices. Their properties are all the same, only the single-

[1]See above, Chapter 3, p. 72.

colored type is light in weight and easily crushed, and that with a black linear pattern is heavy and difficult to crush. There is no difference between all the others.

Grinding malachite is the same as grinding cinnabar or azurite; it is best to grind it dry and to grind it very fine. Before adding glue, you should boil it first and skim off the froth and the grey mud floating on the surface. Wait for it to cool, add glue, and then you must grind it in a mortar, only afterwards proceeding with the liquid decantation. If you use a bamboo tube for natural sinking and floating, this is possible; however, the resulting divisions will have the "green flower" and "branch green" mixed together, and the "third green" and "second green" mixed together, with first green very distinct. My experiment is this: first, the malachite is divided into three colors using the bamboo tube; then using the liquid decantation method separate the "green flower" from the "branch green," the "second green" from the "third green," and the results of doing it this way will be very satisfying. If you work completely by the liquid decantation method, too much time is consumed. After preparing the colors, the glue has to be removed for it to be stored away. However, the glue cannot be entirely removed from the "green flower" and "branch green"; they can also be dried in the sun and stored away without removing the glue.

"First red" produced by a liquid suspension of cinnabar, "first blue" in an azurite suspension, and "first green" in a malachite suspension can all be heat-processed. When heated very hot, take advantage of the heat by plunging them into cold water for a "blast," which will make them break down further. By repeated grinding and liquid suspension of them, they can each be separated into colors of three different shades. The ancient painters and folk artisans all said, when using a mortar to grind up mineral pigments, if you begin by turning the pestle toward the left then you must continue turning it toward the left all along until the pigment is ground up fine, and if you begin by turning the pestle to the right then it must be rotated to the right from start to finish; only like this can the pigment be ground up fine. If it is ground first left and then right, then the pigment cannot be ground up as fine as dust, for if it has been ground by turning sometimes to the right and sometimes to the left then the grains of the pigment will roll up into little round beads. The small beads are round and slippery and they will not stay still in the mortar to be ground up, so naturally they cannot be ground up fine. However, under a magnifying glass we can examine flower petals painted with "red standard" and bamboo leaves painted with "green flower" ("red standard" is the finest portion of cinnabar, while "green flower" is the finest part of malachite) and see small granules adhering to the paper or silk, but looking with the naked eye the application of the color will seem truly liquid, extremely smooth and even, and the granulation cannot be seen. From this it can be seen that if you plan to pound the pigment extremely fine, you cannot ignore that the direction of the hand when grinding should revolve to the left from start to finish or from start to finish toward the right. Only by applying the effects of the principles of physical decomposition, heating it very hot and blasting it with cold, can the problem of coarse cinnabar, coarse azurite, and coarse malachite be solved while increasing their ability to "stand out."

7. Indigo

The raw materials for manufacturing indigo are, first, deep indigo and, second, indigo. In the past, because indigo was not easy to find, deep indigo alone was used. Deep indigo can be bought at Chinese pharmacies, although it is rather expensive. At the pharmacies it is called "Jian deep indigo" (Jian refers to Fujian Province.) Its physical quality is very light and it will float on water. Clear glue must be added and then it is pounded fine with a wooden stick. The pounding must be done patiently, going from a watery state, pounding it until it is dry, then from a dry state gradually adding water while pounding it again until it is watery. In this way, pound it back and forth until it is fine, smooth, and moist and the brilliance of the blue has been brought out. Afterwards, remove it and dry it in the sun, then place it in a large bowl and soak it by turns with warm boiled water until no more yellow liquid emerges. Add clear glue and stir it very evenly, then add cooled boiled water. By this point it should make a full bowl of blue liquid. Skim off this blue liquid; drying it in the sun or by heat will do equally well, and this, then, is the indigo that we want. The sediment of the leftover and sunken indigo can be added in and pounded again the next time that indigo is made. Remember to use only boiled water from beginning to end, not a drop of unboiled water, by which you can reduce the curdling of the glue.

The preparation of other colors can be done no matter what the season is (an exception being the springtime in Beijing when there is a lot of wind-blown sand and the dust is thick), but indigo alone can only be made in summer and autumn. On sunny days you can dry it in the sun, and on overcast and rainy days you can bake it dry.

8. Clamshell white

If store-bought clamshell white is ground fine in a mortar with glue added, the glue cannot be extracted. After using it, dry it in the sun; when using it, add water and grind it up again. After it has been used several times, what is stored in the bowl will be finer and smoother. It should not be used up, but you should put in more white, add glue, and grind it up again; in this way, alternately adding liquid and adding material, grinding it for use, using and grinding it again, it will be much better for use.

For red standard, the standard of red ochre, indigo, and carmine, a painting dish can be used to hold them, while for the other colors a small mortar should be used to contain them. The best mortars are the porcelain ones used in Western medicine, while the glass ones should not be used because boiling water must be used to extract the glue. Through usage and grinding, grinding and usage, the amount of color contained increases and becomes increasingly fine. The commonly used plum blossom type, six-cornered type, and other porcelain painting dishes are not very suitable for azurite, malachite, cinnabar, clamshell white, or iron red.

Methods of Application

Although each of the methods by which traditional-style Chinese painters apply color has its variations, still, when you have an essential grasp of the rules for these techniques their basic principles are the same. The following simply presents several principal colors to serve as examples.

1. The application of cinnabar

Regarding cinnabar, generally, only the two colors, "red standard" and "second red" (pure red), are used. Whenever you apply these colors in landscape painting to depict the red morning sun, colored sunset clouds, or maple leaves, this should only be done with even gradation, followed by a layer of alum and then a layer of carmine wash. However, in figure painting or flower-and-bird painting, white must first be used as a ground (if white is not used for a ground, it will be difficult to apply the washes evenly); then apply a thin wash of "second red," after which apply alum and then a layered wash of carmine, and this will give a feeling of depth. What is most to be avoided is a thick application; furthermore, it is particularly unsuitable to use "first red" to paint the backside. To paint the backside of painting silk, only red standard and white can be used. Also, with large red camellias, peony blossoms, and others, a feeling of substance and volume must be expressed in the painting. The method for the wash is this: first apply a thin wash of white, then use cinnabar from the petal tip and edges toward the inner recesses of the petal in a modulated wash (using two brushes at the same time, one dipped in cinnabar and one dipped in water, the cinnabar brush for the wash and the water brush for modulation), modulating it toward the recesses of the petals, the concavities, the curved and folded areas where the cinnabar color becomes extremely light, and where there is no cinnabar at all. After it has dried, apply alum and then use carmine from the recess of the petals and the concavities, modulating it outwards toward the edges and tips of the petals. (This time use only one brush for the modulated wash, first dipped in clear water and the tip then dipped in carmine or rouge, applying the wash from the base of the petal toward the tip, becoming lighter as it approaches the tip. This method is called an "overlaid wash" and also called a "drawn-out wash.") If one layer is not enough, add another until the flower petals have the feeling of being suspended in space, flying and dancing in the wind. In their use of modulated washes of cinnabar, the early Chinese painters all relied on rouge. Washed on in much quantity, rouge becomes dark and is not as bright and pleasing to the eye as a wash of carmine.

2. The application of carmine

The carmine produced by the Greater German Pigment Company does not change color, while that produced by the Pelikan Company changes color somewhat. The color of the former is a deep red, while the latter by comparison is somewhat more "sharp" and is known as "peony red." (Buy it in cakes with the glue added for use. The Jiang Sixu Pigment Shop still sells a peony red which they make as a paste and which can be used after dissolving it in water.) When these colors are used as layered washes for flowers and fruit, they always are used with a brush tip that has been dipped to go from a deep wash to a pale one. If one layer is insufficient, add a second or third layer; the more washes, the redder it will be. If flower petals are to be a velvety purple or if clothing is to be a magnificent purple, and so forth, it is necessary to first lay down a good ground of light indigo and then to add several layers of wash in carmine, which will make a purple that issues from the red or a red that emerges from the purple, uncommonly beautiful and radiant.

When copying early paintings, for colors that have already darkened, in carrying out the reproduction do not simply add a quick application of dull brown or dark grey color directly over the bright colors, as that will not be sufficient to match the tone of the already darkened color. For instance, the cinnabar skirts in the anonymous Tang painting *Court Ladies with Silk Fans,* formerly displayed in the [Beijing] Palace Museum Hall of Painting, and the cinnabar feathers on the [bird's] chest in Zhao Ji's *Lotus and Golden Pheasant* both have a red that has already darkened. To copy these two early works, first use the best and brightest pure red (second red) applied in very thin washes. Wait for that to dry, then apply a light alum solution over it. When this is thoroughly dry, it will be brighter than the feathers on the chest of the golden pheasant. Next apply a covering wash of rouge with some carmine and red ochre added, and with one or two layers it will naturally become a darkened cinnabar. In addition, to copy the cinnabar skirts the women are wearing [in *Court Ladies*], an iron red covering wash should be applied before the rouge is used, with the result that it darkens to look just like the color of cinnabar that has aged for eighteen hundred years. Thus, by analogy, when using azurite, malachite, or white, always use top quality color to lay down a ground and then, matching the degree of darkening of color in the early painting, proceed to cover it with washes. Under no circumstances should you use the method by which some dark brown or dark grey color is rubbed on in a single layer over a brighter color, especially with white, for the degree of whiteness was different in the ancient paintings during each of the periods, Han, Jin, Tang, and Song. For example, when making a copy of a painting, one cannot simply mix up white with some other light grey color and smear it on and have it be able to stand for the white of the remote eras of Han or Jin, Tang or Song. The first thing to investigate is whether the pigment the ancient master used was white chalk, clamshell white, or whatever, and then determine what to use as a ground as well as what other appropriate greyish colors to use, proceeding to cover it with washes until they match.

3. The application of indigo

The most common use of indigo is for washes on the leaves of flowers and plants.

The boneless method of painting is this: first use a brush dipped in a mixed light grass green color, then dip the tip into a bit of indigo and with a single stroke going from the base of the leaf to the tip you can form a single green leaf with a deep-colored base and a light-colored tip. This is similar to painting pink flower petals where the brush is first dipped in water or a pale white and the tip is dipped in rouge or carmine, with a wash drawn out from the tip of the petal to the base, forming a petal with a red color at the tip that becomes paler down toward the base. Furthermore, taking a brush first dipped in light white then dipped in a bit of light yellow-green [probably meaning sophora yellow made from young buds], with the tip next dipped in carmine or rouge, use it to paint an orchid blossom, first painting the small petals in the heart of the blossom, then painting the outer, larger petals. In this way, you can not only paint orchid blossoms in dark and pale tones but you can paint the red tips of the petals and fine red veins. If the brush is first dipped in light yellow-green [probably sophora yellow], then in indigo, and the tip is next dipped in a little red ochre or red (including rouge or carmine), when used to paint a stroke drawn from the leaf tip to the leaf base, five colors can be painted onto a single leaf. Dipping the brush into red ochre will make old leaves, while dipping it into red will make fresh leaves, bright and colorful, painted in a single stroke and at once rendering the season and time of day. If you dip the brush in indigo and then into carmine to paint the flowers of Chinese wisteria, you can wash on flowers and buds of varying shades and concentration. This kind of painting must be completed in a single stroke, and the thing to be avoided most is going over it repeatedly.

When employing indigo for leaf blades done in outline-and-fill or outline-and-draw (outline-and-fill is filling in color following along ink outlines; outline-and-draw uses color over ink lines, then when the color has been added completely a colored line or an ink line is employed to draw them again), first use layered washes of indigo to differentiate thick and thin, back and front, light and heavy, dark and pale. On top of that, either a covering wash of malachite or one of grass green is acceptable. However, after applying indigo, a layer of alum solution must first be laid down, then a green covering wash, and the indigo will remain fixed and stable. The dark green leaves of osmanthus, camellias, and others require that a little light ink be added to the indigo and then, when applying layered washes, that you use a green covering wash. Some artists put down a layer of grass green covering wash, and if they feel this to be inadequate they can cover it with another layer or two of wash, adding washes until it is verdant, luxuriant, and beautiful, in the spirit of the real object depicted and sometimes perhaps even surpassing it. In addition, if one is painting ink peonies, indigo washes should first be used to distinguish the areas of thick and thin, back and front, light and heavy, dark and pale, and afterwards use more than one layer of carmine as a covering wash. Distant mountains, the color of sky, and other things in landscape paintings also employ shaded washes of indigo.

4. The application of azurite and malachite

In clothing, in the feathers of birds, and in landscape paintings, in mountains, rocks, trees, the color of sky, and certain other objects there are many opportunities to use azurite. If different degrees of concentration and shades are sought on the blue surface, then first a base of ink or indigo must be made to bring forth the thick and thin, dark and pale, after which very thin washes of azurite are applied, and then it will naturally form hues of different concentrations and shades. For both "second blue" and "third blue," always add the glue at the time they are to be used and after their use remove the glue. For the mountains and rocks in landscape paintings, besides first using layered washes of ink or indigo, you also have to apply malachite at the same time in order to avoid the defect of stiffness and lifeless rigidity. As for the applied azurite "ground" (apart from the painted part of the painting, the still-unpainted sections are called the "ground"), "first blue" should be used for this. The method of application is this: to begin with, the "first blue" should be mixed with a glue solution and you must figure out in advance the amount to be used, too much being better than too little. In the south, on winter days there is no fire indoors, so in all seasons a "ground" can be applied. During winter in the north, indoors there is either a fire going or a heating pipe, so one must take some prepared old newspapers, having sprinkled water on them in anticipation of their use, and then proceed with the application. After painting one section completely, use the damp paper to cover the surface (do not let the damp paper actually come into contact with the areas where the blue has been applied), covering it over as it is applied and continuing until the application is complete. Afterwards, remove all of the damp paper so that it can all dry at the same time, and in this way the entire piece can be even and absolutely free of brushmarks. The rest of the time, the more the weather is overcast and rainy, the easier it will be to apply the color evenly. In adding glue, the consistency must be right because if there is too much glue the pigment will crack easily, while if there is too little glue it will not be easy to make it even, so you must first gain some experience with it, then apply it, and then it will work. The main thing to avoid is applying a ground of azurite in very windy or hot, dry weather.

The uses of malachite are rather numerous. In the blue-and-green mountains and rocks of landscape painting, whether on paper or silk, you should first use ink for the texture strokes and washes to indicate shade and sunlight, front and back, thick and thin, light and heavy, applying these until, as it is said, "There being enough ink resonance, you can fill in the colors." However, when the ink texture strokes and washes are sufficient, first an overall layer of covering wash in light red ochre must be applied (even the sections prepared for the application of the malachite are included in this). After it has dried, first use prepared "third green" well mixed with sophora yellow, covering the areas where you think malachite ought to be used with a very thin wash, making it lighter in those areas toward the foot of the rocks, the base of mountains, and areas of mist and clouds, to truly resemble the scene depicted in an ancient poem: "The colors of the mountains pale into nothingness." In this way, put down one layer of wash at a time until you consider the color sufficient. Even if you are painting a green-and-gold landscape using pure blue and pure green, "third green" and "second green" should be intermixed in the application, gradually sparing out the washes toward the base of the mountains and rocks so that they expose the color of the red ochre washes. On paper, after applying this kind of a covering wash, because the application of malachite is so thin, the outlines, texture strokes, and washes of the mountains will give the impression that it is the malachite that forms the shade and sunlight, front and back, thick and thin, dark and pale. After it has dried thoroughly, then use a modulated wash of grass green, which will add to the clarity and depth. On silk, "first green" should be used to paint the backside. When using a very thin covering wash of malachite, if you still feel that the color is insufficient, you can add another covering wash of malachite. Before adding this malachite covering wash, a layer of light alum solution must first be put down. Wait for it to dry and then apply the covering wash. Even if the malachite covering wash is sufficient, before applying the modulated washes of grass green, a layer of light alum solution should also be applied. After the alum has dried thoroughly, press the painting down flat and wipe it with a cloth to see if the color will hold fast or not. If it does not hold, repair it without delay; if it has held, then proceed with the grass-green modulated wash. In this way, in order not to have the malachite spread as soon as the shading washes are added—the technical term in coloring is "rolling"—when the application of malachite is completed, add a layer of light alum solution and wipe it with a cloth in order to increase its adhesive strength. It is the same for landscape painting as well as for other kinds of painting—after completing the application of malachite, set down a layer of alum and wipe it with a clean cloth.

To use malachite for the leaves of flowers and plants, first use a layered wash of indigo to bring forth the thick and thin, dark and pale, light and heavy places in the leaves, and after that has dried cover it with a layer of alum. Then cover it with a wash of malachite going from the tips and edges of the leaves to the base. The green of the tips and edges of the leaves should be a little darker; in the areas where the indigo is heavy, the malachite covering wash should be thin, so that in the areas washed in with indigo, the malachite will take on a little heavier form. When the washes are finished, wait for them to dry, then put down a layer of alum solution, after which apply a covering wash of grass green. Some leaves are a deep green, while some, such as lotus leaves and others, are suffused with a blue color. These require the use, after a grass-green covering wash, of a very thick covering wash of "third blue" in the direction of the dark, heavy areas. These leaves become all the more dark and deep, suffused with a blue glow. If they are comparatively large leaves, such as lotus leaves, banana leaves, or others, they should even more so be done in this manner in order to give the viewer a sense of three-dimensionality and a sense of realism. On paper there is no need to paint the backside, but on silk the backside must be painted. If wild grasses, flowers, and plants are painted on silk, use a layered light indigo wash first for the green leaves on the front, then a grass-green covering wash. The backside is then painted with malachite to make the colors more fresh and bright. The green color on the backs of the leaves should be comparatively light; do not use indigo as a first wash for them but a "green flower" or "branch green" covering wash, then alum over that and modulated washes of light grass green. On paper, leaves are not painted on the backside, while "third green" mixed with white is used for painting the backside of silk.

Azurite and malachite should both be used in painting peacocks, parrots, and other birds. Because azurite and malachite are relatively opaque, painters take advantage of this. First they put down an ink ground of varied shades and depth, then they use azurite and malachite as covering washes on top of this, which very naturally brings out variations in their depth and tone. In painting birds, when the covering washes of blue and green are sufficient and after applying alum, wipe the surface with a cloth and then apply layered washes of indigo or grass green. Last of all, "thread" the feathers—"threading" is a technical term that means to paint in fine lines.

In the Song Dynasty, there were paintings of azurite-colored peony flowers—the *Peony Manual* called them "ink dancing blue lions"—that had a ground done first in white with rouge then washed over that, and because the petal tips were covered with a wash in an extremely pretty "Yunnan blue," they displayed a magnificent kingfisher-blue suffused with a purple glow. In addition, green-colored peonies used "branch green" for layered washes over the tips and edges of the petals and then used a white modulated wash, after which a light yellow-green [sophora yellow] was used to "catch the eye." "Eye-catching" means to make the layered washes clearer and more striking to the eye; it is also a technical term.

5. The application of white

Thanks to current scientific progress, titanium white and zinc white can both be obtained, while at the same time clamshell white has also been made available. The painters of the Ming and Qing could never have anticipated this. Titanium white and zinc white that have had glue added and have been ground up fine are excellent to use, being strong in color and relatively opaque. When they are applied as washes, it is easy to make them even. When they are used as a base, all types of colors can be used on top of them as covering washes and they are fully adequate for the task.

In using clamshell white, one must first gain some experience with it for it not to be unevenly light and dark. When it is first applied, its whiteness is not apparent, so you have to wait a half-minute to a full minute while the moisture evaporates or seeps out before the white will appear. Therefore, one must first have acquired some experience with it and grasped how dark or light it will be before going ahead with the application of it as a wash, for only then will one be able to create an even tone. Its white color is not as bright as that of titanium white or zinc white, but its color is profound in its simplicity, while the tone is rather reserved, having its own special emotional appeal.

6. *The application of other colors*

Ochre must be divided into three colors for use. To use rattan yellow, the amount to be used must be planned. If you want to use a little bit more, at the time it is needed grind and use the rattan yellow as if it were ink. When the grinding is done, you must wipe the cake of rattan yellow dry. If you want to use a little less, apply the brush to the stick to lap up some of the rattan yellow. The thing to avoid most is soaking it in water, especially in hot water. If this happens once, the color will lose its bright freshness; twice, and the rattan yellow will easily turn red and become hard. For mineral yellow, use that which is manufactured in France, divided into three layers for use. If a ground is laid down first, clamshell white must be used and lead white must be avoided.

Among foreign colors, there is a chemical product called "heliotrope" which resembles the purple color of Chinese redbud blossoms. Bird-and-flower painters often use it mixed together with carmine. It is used for the petals of lotus blossoms and for feathers on the backs of mandarin ducks. As for the combining of different colors, Chinese painters have become increasingly fond of using the expressive vitality of colors. As for mixing colors according to inflexible rules of proportion, as there is not much use in this there is no need to take up space with it here.

Finally, as for this book as a whole, it is merely some very immature and meager research material. On the one hand, I hope that our artists will continue the fine tradition of the early painters' use of color; on the other hand, I seek for our painters, under the brilliant illumination of this country's general policy line for the transition period [put forth by Mao Zedong in 1953], to rely on unceasing practice and on experimentation that is increasingly progressive and scientifically based, while at the same time enriching ourselves by assimilating foreign nutrients, so that our nation's painting will attain even greater achievements. That is the motive and desire of the author in writing this book.

Appendix

Sources for Color Illustrations of Paintings Cited in the Text

In section A, paintings are listed alphabetically by artist or site, together with the present location of the work (if known) and sources where they are illustrated. Full bibliographic information on these sources is given in section B.

A. Major Paintings Cited in the Text

Changsha, Chenjiadashan, Warring States period funeral banner: Woman, dragon, and phoenix. Hunan Provincial Museum, Changsha. *Zhongguo wenwu*, 3 (September 1980): 31; *Chūgoku hakubutsukan*, vol. 2, pl. 52.

Changsha, Mawangdui, Han Tomb no. 1, funeral banner. Hunan Provincial Museum, Changsha. *Changsha Mawangdui yi hao Han mu*, vol. 2, pls. 71–77; *Xi Han bo hua*, 12 colorplates; *Zhongguo wenwu*, 4 (October 1980): 43; *Chūgoku hakubutsukan*, vol. 2, pls. 79–80.

Changsha, Zidanku Tomb no. 1, Warring States period funeral banner: Man riding a dragon. Hunan Provincial Museum, Changsha. *Changsha Chu mu bo hua*; *Zhongguo wenwu*, vol. 2, (September 1980): 32; *Chūgoku hakubutsukan*, vol. 2, pl. 53.

Changsha, Zidanku Tomb no. 1, Chu Silk Manuscript, Arthur M. Sackler Gallery, Washington, D.C. Nakata, *Chinese Calligraphy*, pl. 2.

Dizhangwan (Xianyang) Tang Dynasty tomb. Murals from the tomb of Lady Xue (d. 710). Apparently, now in the Museum of Chinese History, Beijing. Akiyama, *Arts of China*, vol. 1, pl. 197 (detail).

Dong Yuan. *The Xiao and Xiang Rivers*. Palace Museum, Beijing. Weng and Yang, *Treasures of the Forbidden City*, pl. 82; *Zhongguo lidai huihua*, vol. 1, pls. 98–100.

Dunhuang cave paintings. Numerous volumes have appeared recently, one of the most accessible being Dunhuang Institute, *The Art Treasures of Dunhuang*.

Gu Hongzhong. *The Night Entertainment of Han Xizai*. Palace Museum, Beijing. *Zhongguo lidai huihua*, vol. 1, pls. 84–93; Weng and Yang, *Treasures of the Forbidden City*, pl. 83a–b (two details).

Gu Kaizhi. *Admonitions of the Court Instructress*. British Museum, London. *National*

Palace Museum Quarterly, 7, no. 3 (Spring 1962), pl. 4; Cahill, *Chinese Painting*, p. 13 (detail).

Han Huang. *Scholars in a Garden*. Palace Museum, Beijing. *Yi yuan duo ying*, 1979.1, pl. 1; *Zhongguo lidai huihua*, vol. 1, pls. 44–45.

Huang Quan. *Rare Birds Painted from Life*. Palace Museum, Beijing. Weng and Yang, *Treasures of the Forbidden City*, pl. 81; *Zhongguo lidai huiha*, vol. 1, pls. 74–75.

Lang Shining (Giuseppe Castiglione). Various paintings by this artist are illustrated in color in Beurdeley, *Giuseppe Castiglione*.

Liaoyang Han Dynasty tomb (Beiyuan). Poor quality color illustration in *Wenwu cankao ziliao*, 1955.5, frontispiece.

Linyi, Jinqueshan, Han Tomb no. 9, funeral banner. *Wenwu*, 1977.11, colorplate 1 (detail); *Zhongguo wenwu*, 4 (October 1980), p. 43 (two details).

Lu Lengjia. *Six Venerables* or *Buddhist Lohans with Worshipers*. Palace Museum, Beijing. *Zhongguo lidai huihua*, vol. 1, pls. 58–69.

Wang Ximeng. *A Thousand Leagues of Rivers and Mountains*. Palace Museum, Beijing. *Wenwu*, 1979.2, frontispiece (detail); Weng and Yang, *Treasures of the Forbidden City*, pls. 88a–b (two details); *Zhongguo lidai huihua*, vol. 2, pls. 94–132.

Wang Wei. *Wang River Villa*. Various later versions are reproduced in Kohara, *Ō I*.

Wangdu Han Dynasty tomb murals. *Han Tang bihua*, pls. 8, 10, 13, 15, 16.

Xianyang Qin Dynasty Palace no. 3, mural. *Kaogu yu wenwu*, 1980.2, colorplate 2.

Yan Liben, attributed. *Emperors of Successive Dynasties* or *Thirteen Emperors Scroll*. Museum of Fine Arts, Boston. Fontein, *Museum of Fine Arts, Boston*, pls. 67–69 (three details).

Zhan Ziqian. *Traveling in Springtime*. Palace Museum, Beijing. Weng and Yang, *Treasures of the Forbidden City*, pl. 78; *Zhongguo lidai huihua*, vol. 1, pls. 33–35.

Zhang Xuan, copied by Zhao Ji (Song Emperor Huizong). *Lady Guoguo on an Outing*. Liaoning Provincial Museum. *Chūgoku hakubutsukan*, vol. 3, pls. 93–94. For a different version of this work, attributed to Li Gonglin, see Cahill, *Chinese Painting*, p. 20.

Zhao Gan. *Early Snow Along the River*. National Palace Museum, Taibei. Cahill, *Chinese Painting*, p. 21 (detail).

Zhao Ji. *Rare Birds Painted from Life*. See Huang Quan.

Zhao Ji (Song Emperor Huizong). *Listening to the Zither*. Palace Museum, Beijing. Fourcade, *Art Treasures of Peking*, pl. 21; *Zhongguo lidai huihua*, vol. 2, pls. 92–93.

Zhao Ji. *Lotus and Golden Pheasant*. Palace Museum, Beijing. *Gugong bowuyuan hua niao hua xuan*, pl. 2; *Yi yuan duo ying*, 1979.1, pl. 23.

Zhao Ji. *Finches and Bamboo*. Metropolitan Museum of Art, John M. Crawford, Jr., Collection. Cahill, *Chinese Painting*, p. 73.

Zhou Fang. *Court Ladies Wearing Flowers*. Liaoning Provincial Museum, Shenyang. Akiyama, *Arts of China*, pl. 199a–b (two details); *Chūgoku hakubutsukan*, vol. 3, pls. 84–86.

B. Bibliographic Sources

Akiyama, Terukazu, et al. *Arts of China, I: Neolithic Cultures to the T'ang Dynasty, Recent Discoveries.* Tokyo: Kodansha, 1968.

Beurdeley, Cecile, and Michel Beurdeley. *Giuseppe Castiglione: A Jesuit Painter at the Court of the Chinese Emperors.* Rutland, Vermont: Charles Tuttle, 1971.

Changsha Chu mu bo hua (Silk Painting from the Chu Tomb at Changsha) 長沙楚墓帛畫. Beijing: Wenwu chuban she, 1973.

Chūgoku hakubutsukan (Chinese Museums) 中國博物館. Vol. 2, Hunan Provincial Museum, and Vol. 3, Liaoning Provincial Museum. Tokyo: Kodansha, 1981–82.

Dunhuang Institute for Cultural Relics. *The Art Treasures of Dunhuang: Ten Centuries of Chinese Art From the Mogao Grottoes.* Hong Kong: Joint Publishing Company, 1981.

Cahill, James. *Chinese Painting.* Lucerne: Skira, 1960.

Fontein, Jan, and Pratapaditya Pal. *Museum of Fine Arts, Boston: Oriental Art.* Greenwich, Connecticut: New York Graphic Society, 1969.

Fourcade, Francois. *The Art Treasures of Peking.* New York: Harry Abrams, 1965.

Gugong bowuyuan cang hua niao hua xuan (The Palace Museum Collection of Flower-and-Bird Paintings) 故宮博物院藏花鳥畫選. Beijing: Wenwu chuban she, 1965.

Han Tang bihua (Han and Tang Wall Paintings) 漢唐壁畫. Beijing: Wenwu chuban she, 1974.

Kaogu yu wenwu (Archaeology and Cultural Relics) 考古與文物. Xi'an.

Kohara, Hironobu 古原宏伸. *Bunjinga suihen.* Vol. 1, О I 文人畫粹編：王維. Tokyo: Chūsōkōronsha, 1975.

Nakata, Yūjirō, editor. *History of the Art of China.* Vol. 1, *Chinese Calligraphy.* New York: Weatherhill/Tankosha, 1983.

Mawangdui yi hao Han mu (Han Tomb No. 1 at Mawangdui) 馬王堆一號漢墓, 2 vols. Beijing: Wenwu chuban she, 1973.

Wenwu (Cultural Relics) 文物. Beijing.

Wenwu cankao ziliao (Reference Materials on Cultural Relics) 文物參考資料. Beijing.

Weng, Wan-go and Yang Boda. *The Palace Museum, Peking: Treasures of the Forbidden City.* New York: Harry Abrams, 1982.

Xi Han bo hua (Western Han Silk Painting) 西漢帛畫. Beijing: Wenwu chuban she, 1972.

Yi yuan duo ying (A Collection of Fine Arts) 藝苑掇英. Beijing.

Zhongguo lidai huihua: Gugong bowuyuan cang hua ji (The History of Chinese Painting: Paintings from the Palace Museum Collection 中國歷代繪畫：故宮博物院藏畫集. 4 vols. Beijing: Wenwu chuban she, 1978–84.

Zhongguo wenwu 中國文物. Beijing.

Selected Bibliography

Acker, William. *Some T'ang and Pre-T'ang Texts on Chinese Painting*, 2 vols. Leyden: E. J. Brill, 1954, 1974.

Fong, Mary. "The Technique of 'Chiaroscuro' in Chinese Painting From Han Through T'ang." *Artibus Asiae*, 38, nos. 2–3 (1976).

Franke, Herbert. "Kulturgeschichtliches über die chinesiche Tusche." *Abhandlungen, Bayerische Akademie der Wissenschaften*, 54 (1962).

————. "Two Yüan Treatises on the Technique of Portrait Painting." *Oriental Art*, 3 (1950).

Gettens, Rutherford J. "Calcium Carbonate Whites." *Studies in Conservation*, 19, no. 3 (August 1974).

Gettens, Rutherford J., Robert L. Feller, and W. T. Chase. "Vermilion and Cinnabar." *Studies in Conservation*, 17, no. 2 (May 1972).

Gettens, Rutherford J., and Elizabeth W. Fitzhugh. "Azurite and Blue Verditer." *Studies in Conservation*, 11, no. 1 (May 1966).

————. "Malachite and Green Verditer." *Studies in Conservation*, 19, no. 1 (May 1974).

Gettens, Rutherford J., Hermann Kühn, and W. T. Chase. "Lead White." *Studies in Conservation*, 12, no. 4 (November 1967).

Gettens, Rutherford J., and George L. Stout. *Painting Materials: A Short Encyclopedia*. New York: Van Nostrand, 1942.

Glum, Peter. "Light Without Shade: The Divine Radiance, Moonlight, Sunsets, Translucence, Luster and Other Light Effects in Chinese and Japanese Painting." *Oriental Art*, N.S. 27, no. 4 (Winter 1981–82), N.S. 18, no. 1 (Spring 1982).

————. "Meditations on a Black Sun: Speculations on Illusionistic Tendencies in T'ang Painting Based on Chemical Changes in Pigments." *Artibus Asiae*, 37, nos. 1–2 (1975).

Hansford, Howard. *A Glossary of Chinese Art and Archaeology*. London: China Society, 1954.

Ippolito, Jean. "Chinese Paint Pigments and Their Classification, with an Examination of Four Ming Paintings at the Seattle Art Museum." M.A. Thesis, University of Washington, 1985.

Kexue mingci huidian (Dictionary of Scientific Words and Terms) 科學名辭滙典. Vols. 4 and 5. Taibei: Zheng zhong shuju, 1959.

Kühn, Hermann. "Verdigris and Copper Resinate." *Studies in Conservation*, 15, no. 1 (February 1970).

Laufer, Berthold. "The History of Ink in China." In Frank Wiborg, *Printing Ink: A History*. New York: Harper and Brothers, 1926.

Li Ch'iao-p'ing. *The Chemical Arts of Old China*. Easton, Pennsylvania: Journal of Chemical Education, 1948.

Li Kan 李衎. *Zhu pu (Bamboo Manual)* 竹譜. *Hua lun congkan* edition 畫論叢刊. Vol. 2. Taibei: Zhonghua shuju, 1977.

Lu Hongnian 陸鴻年. "Zhongguo hua bi zhifa diandi (Notes on the Methods of Painting Walls in China)" 中國畫壁制法點滴. *Wenwu* 文物, 1956.8.

March, Benjamin. *Some Technical Terms of Chinese Painting*. Baltimore: Waverly Press, 1935.

Mayer, Ralph. *The Artist's Handbook of Materials and Techniques*, 4th ed. New York: Viking Press, 1981.

————. *A Dictionary of Art Terms and Techniques*. New York: Apollo, 1969.

Mühlethalter, Bruno and Jean Thissen. "Smalt." *Studies in Conservation*, 14, no. 2 (May 1969).

Needham, Joseph. *Science and Civilization in China*. Vol. 5. London: Cambridge University Press, 1974.

Plesters, Joyce. "Ultramarine Blue, Natural and Artificial." *Studies in Conservation*, 11, no. 2 (May 1966).

Read, Bernard E. *Chinese Medicinal Plants From the Pen Ts'ao Kang Mu, A.D. 1596: A Botanical, Chemical, and Pharmacological Reference List*, 3rd ed. Beijing: Peking Natural History Bulletin, 1936.

Read, Bernard E., and C. Pak. *A Compendium of Minerals and Stones Used in Chinese Medicine, From the Pen Ts'ao Kang Mu*. Beijing: Peking Society of Natural History Bulletin, 1928.

Shen Zongqian 沈宗騫. *Jiezhou xue hua bian (Jiezhou's Painting Studies)* 芥舟學畫編. *Hua lun congkan* edition. Vol. 2. Taibei: Zhonghua shuju, 1977.

Song Yingxing 宋應星. *Tian gong kai wu (On Using the Products of Nature)* 天工開物. Beijing: Zhonghua shuju, 1959.

————, translated by E-tu Zen Sun and Shiou-chuan Sun. *T'ien-kung k'ai-wu: Chinese Technology in the Seventeenth Century*. University Park: Pennsylvania State University Press, 1966.

Takamatsu, Toyokichi. *On Japanese Pigments*. Tokyo: Tokyo Daigaku, Department of Science, 1878.

Toda, Teisuke, "The Use of Gold in Southern Sung Academic Painting." Paper presented at symposium on "Words and Images: Chinese Poetry, Calligraphy, and Painting." Metropolitan Museum of Art, 1985. Publication forthcoming.

Uyemura, Rokurō. "Studies on the Ancient Pigments in Japan." *Eastern Art*, 3 (1931).

————上村大郎. *Toho zenshoku bunka no kenkyu (Research on the Art of Dyes and Pigments in the Orient)* 東方染色文化の研究. Tokyo: Daiichi Shobō, 1933.

Wang Gai, et al., translated by Mai-mai Sze. *The Mustard Seed Garden Manual of Painting*. Princeton: Princeton University Press, 1963.

Wang Yi 王繹. *Xie xiang bijue (Secrets of Portrait Painting)* 寫像秘訣 and *Cai hui fa (Technique of Painting in Colors)* 采繪法. *Hua lun congkan* edition. Vol. 2. Taibei: Zhonghua shuju, 1977.

Wen Jinyang 文金楊. *Huihua secai xue (Studies on Painting Colors)* 繪畫色彩學. Jinan: Shandong renmin chuban she, 1982.

Winter, John. "'Lead White' in Japanese Paintings." *Studies in Conservation*, 26, no. 3 (August 1981).

Yamasaki, Kazuo. "Chemical Studies on the Eighth-Century Red Lead Preserved in the Shoso-in at Nara." *Studies in Conservation*, 4, no. 1 (February 1959).

—————山崎一雄. "The Chemical Studies on the Pigments used in the Wall Paintings of the Main Hall of the Hōryūji and their Color Changes by the Fire of January 1949." *Bijutsu kenkyu*美術研究, 167 (January 1953).

—————. "Pigments Used in the Wall-paintings of the Five-Story Pagoda at Daigo-ji Monastery." *Bijutsu kenkyu*美術研究, 196 (January 1958).

—————. "Pigments in the Wall-Paintings in Central Asia." *Bijutsu kenkyu* 美術研究, 212 (September 1960).

Yamasaki, Kazuo, and Yoshimichi Emoto. "Pigments Used on Japanese Paintings from the Protohistoric Period through the 17th Century." *Ars Orientalis*, 11 (1979).

Zou Yigui鄒一桂. *Xiaoshan hua pu (Xiaoshan's Painting Manual)*小山畫譜. *Hua lun congkan* edition. Vol. 2. Taibei: Zhonghua shuju, 1977.

Index